Radiochemical
Methods

Analytical
Chemistry
by Open Learning

Titles in Series:

Radiochemical Methods

Analytical Chemistry by Open Learning

Author:
WILLIAM J. GEARY
Sheffield City Polytechnic, UK

Editor:
ARTHUR M. JAMES

on behalf of ACOL

Published on behalf of ACOL, London
by
JOHN WILEY & SONS
Chichester · New York · Brisbane · Toronto · Singapore

Library of Congress Cataloging in Publication Data:

Geary, William J.
  Radiochemical methods.
  Bibliography: p.
  1. Radiochemical analysis.   1. James, Arthur M.
  II. Title.
  QD605.G43   1986   543'.088   86-9153

  ISBN 0 471 91117 8 (cloth)
  ISBN 0 471 91118 6 (paper)

British Library Cataloguing in Publication Data:

Geary, Willliam J.
  Radiochemical methods. — (Analytical chemistry
  by open learning).
  1. Radiochemical analysis
  1. Title   II. James, Arthur M.   III. Series
  543'.088   QD605

  ISBN 0 471 91117 8 (cloth)
  ISBN 0 471 91118 6 (paper)

Printed and bound in Great Britain

Analytical Chemistry

This series of texts is a result of an initiative by the Committee of Heads of Polytechnic Chemistry Departments in the United Kingdom. A project team based at Thames Polytechnic using funds available from the Manpower Services Commission 'Open Tech' Project have organised and managed the development of the material suitable for use by 'Distance Learners'. The contents of the various units have been identified, planned and written almost exclusively by groups of polytechnic staff, who are both expert in the subject area and are currently teaching in analytical chemistry.

The texts are for those interested in the basics of analytical chemistry and instrumental techniques who wish to study in a more flexible way than traditional institute attendance or to augment such attendance. A series of these units may be used by those undertaking courses leading to BTEC (levels IV and V), Royal Society of Chemistry (Certificates of Applied Chemistry) or other qualifications. The level is thus that of Senior Technician.

It is emphasised however that whilst the theoretical aspects of analytical chemistry can be studied in this way there is no substitute for the laboratory to learn the associated practical skills. In the U.K. there are nominated Polytechnics, Colleges and other Institutions who offer tutorial and practical support to achieve the practical objectives identified within each text. It is expected that many institutions worldwide will also provide such support.

The project will continue at Thames Polytechnic to support these 'Open Learning Texts', to continually refresh and update the material and to extend its coverage.

Further information about nominated support centres, the material or open learning techniques may be obtained from the project office at Thames Polytechnic, ACOL, Wellington St., Woolwich, London, SE18 6PF.

How to Use an
Open Learning Text

Open learning texts are designed as a convenient and flexible way of studying for people who, for a variety of reasons cannot use conventional education courses. You will learn from this text the principles of one subject in Analytical Chemistry, but only by putting this knowledge into practice, under professional supervision, will you gain a full understanding of the analytical techniques described.

To achieve the full benefit from an open learning text you need to plan your place and time of study.

- Find the most suitable place to study where you can work without disturbance.

- If you have a tutor supervising your study discuss with him, or her, the date by which you should have completed this text.

- Some people study perfectly well in irregular bursts, however most students find that setting aside a certain number of hours each day is the most satisfactory method. It is for you to decide which pattern of study suits you best.

- If you decide to study for several hours at once, take short breaks of five or ten minutes every half hour or so. You will find that this method maintains a higher overall level of concentration.

Before you begin a detailed reading of the text, familiarise yourself with the general layout of the material. Have a look at the course contents list at the front of the book and flip through the pages to get a general impression of the way the subject is dealt with. You will find that there is space on the pages to make comments alongside the text as you study—your own notes for highlighting points that you feel are particularly important. Indicate in the margin the points you would like to discuss further with a tutor or fellow student. When you come to revise, these personal study notes will be very useful.

∏   When you find a paragraph in the text marked with a symbol such as is shown here, this is where you get involved. At this point you are directed to do things: draw graphs, answer questions, perform calculations, etc. Do make an attempt at these activities. If necessary cover the succeeding response with a piece of paper until you are ready to read on. This is an opportunity for you to learn by participating in the subject and although the text continues by discussing your response, there is no better way to learn than by working things out for yourself.

We have introduced self assessment questions (SAQ) at appropriate places in the text. These SAQs provide for you a way of finding out if you understand what you have just been studying. There is space on the page for your answer and for any comments you want to add after reading the author's response. You will find the author's response to each SAQ at the end of the text. Compare what you have written with the response provided and read the discussion and advice.

At intervals in the text you will find a Summary and List of Objectives. The Summary will emphasise the important points covered by the material you have just read and the Objectives will give you a checklist of tasks you should then be able to achieve.

You can revise the Unit, perhaps for a formal examination, by re-reading the Summary and the Objectives, and by working through some of the SAQs. This should quickly alert you to areas of the text that need further study.

At the end of the book you will find for reference lists of commonly used scientific symbols and values, units of measurement and also a periodic table.

Contents

Study Guide

This Unit is designed to give you a good background in the theory and practice of radiochemistry as it is applied in analytical chemistry, and to explain in more detail the important methods of radioanalytical determination.

Most analytical chemists are aware that radioactive isotopes can be useful in analytical processes; some are aware of the increasing need to measure quantititively radioactive isotopes in environmental and other samples. However, very few analysts have regular working experience in radiochemistry. This Unit aims to provide information in sufficient detail for non-specialists to understand how radiochemistry can be helpful to the analyst. It also aims to give an appreciation of the value and current position of radioanalytical methods.

To achieve these aims the Unit firstly surveys the basic principles of nuclear chemistry, with particular reference to answering such questions as 'why is a nucleus radioactive?', 'what are the differences between the various types of decay?', 'how quickly does a particular isotope decay?' and 'how can a particular isotope be measured/counted?' The parts of this Unit that are relevant to this aspect start at quite a basic level, since few chemistry courses cover radioactivity in any detail; you really need only a reasonable knowledge of the structure of the atom, and some fairly basic mathematical background. If you already have a good background in radiochemistry the material may be helpful in a revision sense, and may perhaps give you a different perspective on your present knowledge.

Please be clear that the treatment in no way is intended to be a comprehensive and detailed review of radioactivity and its applications. We are interested specifically in applications in analytical chemistry, and so much of the detail of nuclear physics (interesting as it may be) is not included.

The later parts of the Unit are designed to give a good working knowledge of radioanalytical methods from both a theoretical and a practical point of view. Each method is surveyed in turn, and placed in context by consideration of selected published papers.

If you become involved directly in radioactive work you will need more detailed training in certain topics, notably safety procedures and possibly instrumentation. If you are already involved in the field you will hopefully find something new, particularly in the later sections.

Supporting Practical Work

1. GENERAL CONSIDERATIONS

Experimental facilities for radiochemistry are less widely available than for many other analytical techniques, and even where the basic facilities are available the more advanced instrumental equipment may not be. The experiments which follow are each designed to occupy a three hour laboratory period, and to use rather basic counting equipment. If more sophisticated equipment is available, and particularly if (as may well be the case) the learner has a requirement to specialise in a chosen technique, some supplementary experiments are suggested.

2. AIMS

There are four principal aims.

(*a*) To provide basic experience in handling radioisotopes and simple counting apparatus.

(*b*) To illustrate important principles from the theory part of the unit.

(*c*) To illustrate relevant applications of radioisotopes in analytical chemistry.

(*d*) To stress the importance of safe working procedures.

3. SUGGESTED EXPERIMENTS

(*a*) Determination of the response curve of an end-window Geiger counter using a pre-prepared uranium source.

(*b*) Determination of a half-life. For preference this should be the half-life of ^{128}I, following the irradiation of ethyl iodide in an isotopic neutron source. If the latter is not available an experiment for the half-life of ^{234}Pa can be substituted.

(*c*) The use of ^{32}P to determine *either* the efficiency of precipitation of an 'insoluble' phosphate *or* the concentration of phosphate in an unknown mixture.

4.   SUPPLEMENTARY EXPERIMENTS

Depending on the equipment available it is suggested that the following experiments or demonstrations be made available.

(*a*) Use of monitoring equipment for location of contamination.

(*b*) The comparison of NaI(Tl) and Ge(Li) detectors for γ-radiation.

(*c*) The counting of low energy negatron emitters by liquid scintillation methods.

Bibliography

1. 'STANDARD' ANALYTICAL CHEMISTRY TEXTBOOKS

Virtually all the books giving a comprehensive treatment of (instrumental) analytical chemistry contain a chapter on radioactive methods of analysis. Appropriate examples are:

(*a*) H H Bauer, G D Christian and J E O'Reilly, *Instrumental Analysis*, Allyn and Bacon, 1978.

(*b*) F W Fifield and D Kealey, *Principles and Practice of Analytical Chemistry*, International Textbook Co Ltd, 2nd Edn, 1983.

(*c*) H H Willard, L L Merritt, J A Dean and F A Settle, *Instrumental Methods of Analysis*, Van Nostrand, 1981.

2. RADIOCHEMISTRY TEXTBOOKS

(*a*) G R Choppin and J Rydberg, *Nuclear Chemistry, Theory and Applications*, Pergamon, 1980.

(*b*) D J Malcolme-Lawes, *Introduction to Radiochemistry*, Macmillan 1979.

(*c*) D I Coomber, *Radiochemical Methods in Analysis*, Plenum 1975.

(*d*) R A Faires and G G J Boswell, *Radioisotope Laboratory Techniques*, Butterworth, 1981.

(*e*) T A H Peacocke, *Radiochemistry: Theory and Experiment*, Wykeham, 1978.

NOTES:

Reference (*a*) is very detailed and of Honours level and beyond.

Reference (*b*) is more appropriate to this Unit.

Reference (*c*) is the most applicable on analytical methods.

References (*d*) and (*e*) are more directly relevant to the experimental aspects of the technique.

Acknowledgements

Figure 3.3c reprinted with permission from G. R. Choppin and J. Rydberg, *Nuclear Chemistry*, Copyright 1980, Pergamon Press.

'Arsenic and antimony in laundry aids by instrumental neutron activation analysis', J. T. Tanner, M. H. Friedman and G. F. Holloway, *Analytica Chimica Acta*, **66**, 456–459, (1973), copyright 1973, reproduced by permission of Elsevier Scientific Publishing Company, Amsterdam.

'Direct radioimmunoassay for the detection of barbiturates in blood and urine', P. A. Mason, B. Law, K. Pocock and A. C. Moffat, *Analyst*, **107**, 629–633, (1982), Crown copyright 1982, reproduced by permission of HMSO.

'Determination of rhenium by substoichiometric pseudoisotopic dilution analysis with technetium-99 and liquid scintillation counting', R. A. Pacer and S. M. Benecke, reprinted with permission from *Analytical Chemistry*, **53**, 1160–1163, (1981). Copyright 1981 American Chemical Society.

1. Introduction

1.1. NUCLEAR PROPERTIES

Overview

This section is intended to show how a simple treatment of the physical properties of the nucleus, and of the relative numbers of neutrons and protons contained within it, leads to an appreciation of why many nuclei are unstable (ie radioactive).

1.1.1. Physical Properties of the Nucleus

The nucleus of an atom consists of tightly packed nuclear particles and has a very high density. These nuclear particles are often called *nucleons.*

∏ From your previous study of chemistry can you recall an experiment that demonstrates the high density of the nucleus?

One such experiment is the scattering of α particles by thin metal foil.

However, we need a more detailed knowledge of the nucleus and its constituents.

(*a*) The nuclear particles are called either 'neutrons' if they have no electrical charge, or 'protons' if they have a single positive charge.

(*b*) The sum of the protons present gives us the total charge of the nucleus; the number of protons is called the atomic number, and given the symbol Z.

(*c*) The mass of the neutron and the mass of the proton are nearly identical and frequently are quoted as being one 'atomic mass unit', written as 1 amu, 1 m_u, or 1 u. In SI units

$1\ m_u = 1.660\ 566 \times 10^{-27}$ kg, and the exact values of the masses of the nucleons are:

Proton : $1.007\ 276\ m_u = 1.672\ 65 \times 10^{-27}$ kg

Neutron : $1.008\ 665\ m_u = 1.674\ 95 \times 10^{-27}$ kg

(*d*) By comparison, the mass of an electron is vastly smaller:

Electron: $0.000\ 548\ 58\ m_u = 9.109\ 53 \times 10^{-31}$ kg

(*e*) The nucleus of an atom is smaller than the atom as a whole, but can you remember what the relative sizes are? In fact the radius of a nucleus normally is 10^{-4}–10^{-5} of the radius of an atom; hence if the radius of an atom is 10^{-10} m the radius of the nucleus of that atom is likely to be in the range 10^{-14}–10^{-15} m. For this reason, in nuclear physics the unit of nuclear size is often expressed as a 'fermi'; in SI units this is the femtometre, where

1 femtometre (fm) $= 1$ fermi $= 10^{-15}$ m

SAQ 1.1a

> It is instructive to try to relate the radius of the nucleus to that of the atom in more conventional terms. Imagine that the nucleus has a radius of 1 mm. What is the radius of the atom in units of metres?

What conclusions can we draw so far? The principal one is that virtually all of the mass of an atom is concentrated in a nucleus that has a very small radius and carries a positive electrical charge. However, this isn't really enough to support what we know about the highly unusual properties of a nucleus. We have been considering radius and mass; what about *shape, volume,* and *density*? Let us assume that both the atom and its nucleus are spherical; the volume of a sphere is 4/3 (πr^3), so if the radius of the nucleus is 10^{-4}–10^{-5} of that of the atom the volume will be 10^{-12}–10^{-15} of that of the atom ie *vastly smaller*. We can also draw the conclusion that supports the statement with which we started this paragraph: if virtually all the mass of an atom is concentrated in a nucleus of very small volume, the nucleus will have *a very high density*. You can prove this by completing the following exercise:

SAQ 1.1b

(*i*)   Calculate the volume of the nucleus, given that its radius is 10^{-15} m.

(*ii*) If this nucleus contains 20 neutrons and 20 protons each of mass 1 m_u (1 m_u = 1.66 × 10^{-27} kg) calculate the mass of the nucleus.

(*iii*) Hence determine the density of the nucleus.

Isn't this an amazing value? It is far greater than that of even the densest of bulk materials, and should reinforce our conclusion that the structure of the nucleus, and particularly the forces in it, are highly unusual by comparison with those of ordinary matter. For this course we do not need to explore these points further, but if you are interested in pursuing them you will find details in advanced textbooks such as G. R. Choppin and J. Rydberg, *Nuclear Chemistry* (Chapter 6).

1.1.2. Neutron/Proton Ratio

In the previous paragraph we called the number of protons in an atom the atomic number, Z. The total number of protons and neutrons in the atom is called the *mass number* and given the symbol A. Can you remember some of the terminology that follows from these basic definitions? Some important examples are listed.

(*a*) An element is defined by its atomic number, ie by the number of protons in any atom of that element. Thus, for example, all atoms of carbon contain six protons ($Z = 6$).

(*b*) For every chemical element there are atoms with different *mass* numbers. Thus, for example, atoms of carbon can have any mass number from 9 to 16. Obviously, the difference between the atoms of different mass number lies in the number of *neutrons* they contain. An atom of carbon of mass number 9 has 3 neutrons and 6 protons, whereas one with mass number 16 has 10 neutrons and 6 protons. These individual forms of an element, *characterised by the number of neutrons they contain*, are called '*isotopes*' or sometimes '*nuclides*' of the element.

(*c*) If we give a chemical element the symbol X the most precise way of denoting the various isotopes is to write them as $_Z^A X$. Thus the isotope of carbon with 6 neutrons and 6 protons is written as $_6^{12}C$, that with 7 neutrons and 6 protons is $_6^{13}C$, and so on. Sometimes you will see the atomic number omitted (eg ^{12}C), or the name of the isotope written out in full (eg carbon-12).

The next point may seem rather obvious, but it does need saying. Many people equate the words 'isotope' and 'radioactive', simply because *some* isotopes *are* radioactive; don't fall into this trap! Consider some of the isotopes of carbon:

number of protons Z		6	6	6	6
number of neutrons N		5	6	7	8
symbol		^{11}C	^{12}C	^{13}C	^{14}C
stable (S) or radioactive (R)		R	S	S	R

Fig. 1.1a. *Isotopes of carbon*

Is there any obvious reason why two are stable and two are radioactive? What are the variables in Fig. 1.1a? There is only *one* variable—the number of neutrons—and we might therefore suggest that it is this which determines whether an isotope is radioactive or stable. However, if we look more closely we find that it is not as simple as this; for instance, the vast majority (269 of 275) of stable isotopes of the elements have even numbers either of neutrons, or of protons, or both. So shall we now say that even numbers of protons and/or neutrons is a precondition of stability? Again, we cannot make such a simple decision. For example, the isotopes $^{12}_{6}C$, $^{16}_{8}O$, and $^{20}_{10}Ne$ *are* stable, whereas $^{40}_{20}Ca$ is radioactive. Clearly the situation is more complicated, and all that we can safely say at this stage is that the *relative* numbers of neutrons and protons is important. It turns out that we can clarify the problem if, for all stable isotopes, we plot the number of protons (Z) against the number of neutrons (N). Before we draw the graph let us remember that for some elements there is only one stable isotope (and therefore only one value of N relative to Z), whereas for other elements, in fact the majority, there are several. For example, for sodium the only stable isotope is $^{23}_{11}Na$ (ie $N = 12$, $Z = 11$) whereas for chlorine there are $^{35}_{17}Cl$ ($N = 18$, $Z = 17$) and $^{37}_{17}Cl$ ($N = 20$, $Z = 17$). As a consequence we cannot draw a simple graph; most commonly the points for the stable isotopes are plotted in a way which shows them to be in a 'band of stability', represented as the shaded area on Fig. 1.1b

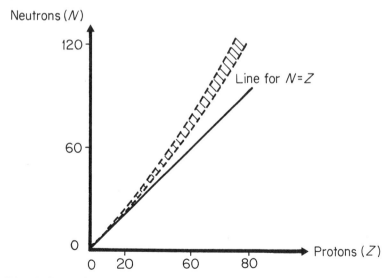

Fig. 1.1b. *Stable nuclei as a function of their proton (Z) and neutron (N) numbers*

What observations can we make?

(*a*) For the lighter elements (roughly to $Z = 20$) the band containing the stable isotopes is more or less superimposed on the line for $N = Z$.

(*b*) Secondly, for elements of $Z > 20$ the 'band of stability' moves increasingly into the 'neutron-rich' section of the graph.

(*c*) Since we have plotted N against Z for the isotopes known to be stable we might anticipate that isotopes lying *outside* this 'band of stability' will be radioactive, and this is true.

We are now getting a little nearer to being able to predict stability or radioactivity for a given isotope, but before we try to be more quantitative we pause and consider some further difficulties. Earlier, we have stated that $^{35}_{17}Cl$ and $^{37}_{17}Cl$ are stable isotopes, and hence the points $N = 18$, $Z = 17$ and $N = 20$, $Z = 17$ lie within the band of stability. *But*, $^{36}_{17}Cl$ is radioactive, even though the point $N = 19$, $Z = 17$ lies within the band of stability. In other words, although we can draw a band of stability that includes all known stable isotopes we cannot say that it totally excludes radioactive isotopes, even though the majority of radioactive isotopes *do* lie outside it.

So in conclusion, can we draw up any empirical rules to predict on the basis of the neutron : proton ratio whether a nucleus will be radioactive?

One simple rule, that should be obvious to you from the graph, is that *all isotopes that have more protons than neutrons are radioactive*; the only exception is the simplest nucleus of all, ^1_1H, which has one proton but no neutrons. For neutron-rich isotopes *no such simple rule exists*, partly because of the way in which the band of stability moves further into the neutron-rich sector of the graph as Z increases. For example, although a particular isotope might have only a slight excess of neutrons over protons (or might even have $N = Z$) it may be radioactive because the $N : Z$ ratio is *lower* than the *minimum* stable value for that element. Thus, both $^{56}_{28}Ni$ and $^{57}_{28}Ni$

are radioactive whereas $^{58}_{28}$Ni is stable. The only real solution is to consult a chart of all known isotopes, but it is *not* realistic to reproduce such a chart here. Certainly it is not satisfactory to set limits on $N:Z$ ratios that apply to all elements. As a generalisation we may say that neutron-rich isotopes will be radioactive if the $N:Z$ ratio exceeds the value in Fig. 1.1c.

	$N:Z$	
Elements below zinc ($Z = 30$)	1.25	(4 exceptions)
Elements from zinc to silver ($Z = 47$)	1.40	(1 exception)
Elements from silver to osmium ($Z = 76$)	1.50	(2 exceptions)

Fig. 1.1c.   *Limiting $N:Z$ values for stability*

SAQ 1.1c

Predict, on the basis of $N:Z$ ratio, whether the following isotopes will be stable or radioactive, by ringing the appropriate responses:

$^{57}_{25}$Mn	S	R
$^{37}_{19}$K	S	R
$^{71}_{31}$Ga	S	R
$^{108}_{44}$Ru	S	R
$^{25}_{12}$Mg	S	R
$^{82}_{38}$Sr	S	R

The main uses of $N : Z$ calculations are in attempting to predict the radioactive isotopes of a given element (if you do not have data available) and, much more importantly—as we shall see in the next section—in relating the value of the ratio to the type of decay process.

1.1.3. Energy

A detailed discussion of the energetics of nuclei is inappropriate for this course, but it is necessary to consider the concept of 'binding energy' since this is often used to assist in the development of theories of nuclear structure and nuclear reactions. Experimentally it is found that the measured mass of an isotope is *less* than the sum of the masses of the constituent particles.

eg: for $^4_2$He, measured mass $\qquad\qquad$ = $\quad$ 4.002 604 m_u

$\quad$ 2 protons + 2 neutrons + 2 electrons $\quad$ = $\quad$ 4.032 980 m_u

$\quad$ Difference (called 'mass defect') $\qquad$ = $\quad$ 0.030 376 m_u

This apparent mass loss is present as energy, (the 'binding energy', E_B) and this can be calculated by using the Einstein mass–energy equivalence $E = mc^2$ where c is the speed of light ($2.997\ 925 \times 10^8$ ms^{-1}).

The most frequently used version of this equivalence is that

$$1\ m_u = 931.5\ \text{MeV}$$

Where 1 MeV = 1 mega-electron volt (10^6 eV).

If you are not familiar with this conversion it can be proved using $E = mc^2$ for a mass of 1 kg:

$$E = 1 \times (2.997\ 925 \times 10^8)^2\ \text{kg m}^2\ \text{s}^{-2}$$

$$= 8.987\ 554 \times 10^{16}\ \text{J} \quad (\text{since 1 J} = 1\ \text{kg m}^2\ \text{s}^{-2})$$

An electron volt is the work done on an electron (charge on the electron = $1.602\ 189 \times 10^{-19}$ C) when moved through a potential difference of 1 volt, ie

$$1\ \text{eV} = 1.602\ 189 \times 10^{-19}\ \text{C V}$$

$$= 1.602\ 189 \times 10^{-19}\ \text{J} \quad (\text{since } 1\ \text{J} = 1\ \text{C V})$$

Therefore in energy terms a mass of 1 kg is equivalent to $8.987\ 554 \times 10^{16}/1.602\ 189 \times 10^{-19}$ eV

$$= 5.609\ 547 \times 10^{35}\ \text{eV}$$

Earlier we said that $1\ \text{m}_u = 1.660\ 566 \times 10^{-27}$ kg

Therefore

$$1\ \text{m}_u \equiv 1.660\ 566 \times 10^{-27} \times 5.609\ 547 \times 10^{35}\ \text{eV}$$

$$\equiv 9.315\ 023 \times 10^8\ \text{eV}$$

$$1\ \text{m}_u \equiv 931.5\ \text{MeV}$$

Hence for ^4_2He $E_B = 0.030376 \times 931.5$

$$= 28.3\ \text{MeV}$$

The binding energy may be looked on as the energy liberated when a nucleus is formed from its component nucleons, or as the minimum energy required to separate all the nucleons in a nucleus and remove them to infinity. Realistically, neither of these processes can be made to occur. What then is the significance of E_B? If E_B is *divided by the total number of nucleons* the value for virtually all nuclei lies in the range 7.5–8.8 MeV, ie is *essentially constant*. (The exceptions are for the very light nuclei of elements up to boron).

This result is of considerable theoretical significance, particularly since it is usually taken to confirm the short range nature of nuclear forces.

SAQ 1.1d

> Calculate E_B for $^{27}_{13}\text{Al}$ given that the measured isotopic mass = 26.981 539 m_u, proton mass = 1.007 276 m_u, neutron mass = 1.008 665 m_u, electron mass = 0.000 548 58 m_u, and confirm that E_B/number of nucleons lies in the expected range.

Finally we should note that in all processes of radioactive decay the parent nucleus is losing some of its energy during its conversion to the more stable daughter nucleus, and it will be necessary to discuss this point further in the next section on types of decay.

SUMMARY AND OBJECTIVES

Summary

The most important physical properties of a nucleus are its small size and very high density. The relative numbers of protons and neutrons are fundamental in determining the stability of a nucleus. The concept of binding energy, particularly in relation to the total number of nucleons, helps us to understand the nature of nuclear forces.

Objectives

You should now be able to:

● appreciate the significance of the small size and high density of the nucleus;

● relate the value of the neutron : proton ratio to the stability (or otherwise) of the nucleus;

● calculate values of binding energy for typical nuclei;

● explain the general significance of nuclear energy changes.

1.2. INSTABILITY AND TYPES OF DECAY

Overview

This section is intended to describe the important properties associated with isotopes which emit α particles, β particles, and γ rays (separately or together). Particular significance is attached to the energy of the emitted particle/ray, to the physical properties: mass (or lack of), charge (or lack of), charge/mass ratio—and to the consequences of these properties in relation to such practical factors as range, penetration, and ease of detection.

The preceding section should have helped you to realise that the most important determinants of nuclear stability are the neutron : proton ratio and the energy of the nucleus. We shall now discuss the various processes by which a nucleus may spontaneously alter its neutron : proton ratio and hence move to a more favourable (ie lower) energy. It is important to realise that we shall concentrate on those properties of the emitted particle or rays that are of practical importance, particularly for analytical purposes. We shall need to recognise also that whilst it is necessary to discuss the different types of decay independently, many nuclei in fact decay by a combination of such events; they are said to have a complicated *decay scheme.*

1.2.1. Alpha Particle Decay

Many isotopes of elements with high atomic number simply contain too many nuclear particles (effectively they are too heavy) to be stable. The most obvious way to remedy this is for the nucleus to eject some of the particles, and the resulting spontaneous emission of two neutrons and two protons together (ie the $^4_2$He nucleus) is known as α particle decay. A typical example is the disintegration of radium.

$$^{226}_{88}\text{Ra} \rightarrow {}^{222}_{86}\text{Rn} + {}^4_2\text{He}$$

Note two important things about the product: firstly it is an isotope of a different element, and secondly it has a slightly *higher* neutron : proton ratio than the parent isotope. These facts suggest that either the parent, or the product, or both, may undergo further decay processes, and there are four such series of inter-related isotopes known. Three of these series start from naturally occurring isotopes (uranium, actinium, thorium) and one from an isotope of the synthetic element neptunium. The complexity of these series (there are 17 radioactive isotopes of 10 elements in the uranium series) poses some experimental problems if individual elements or isotopes need to be measured by following their decay process.

From the viewpoint of their practical applications the important properties of α emitting isotopes are associated with the energy of the emitted particle. There is a favourable energy change associated with α emission ($E_{daughter} < E_{parent}$), and the bulk of the emitted energy is carried away by the α particle. It can be shown that the initial energy of the α particles emitted by a given isotope is effectively constant, and *characteristic of that isotope*; thus, if it is possible to set up an instrument which can measure just those α particles of a selected energy (the technique is called α spectrometry) we have a potential analytical method. This is a developing area of analytical chemistry, particularly for samples containing some of the heavier synthetic elements (eg plutonium) produced through the nuclear power programme.

The other important consequences of energy changes in α emission derive from the fact that although the initial α energy is quite high (typically 5–9 MeV) this is very rapidly lost because the emitted α particle causes intense ionisation in the surrounding medium. Even in air, the range of an α particle is only a few centimetres, and in denser material it is insignificant. Experimentally this causes difficulty in measuring α emitting isotopes by conventional counting equipment (eg Geiger counters), and for this reason such isotopes are very little used as tracers for following chemical processes.

This low penetrating power also means that α emitting isotopes present little external health hazard. However, if they are taken into the body in any way they can cause very severe problems, and for this reason elements such as plutonium are a much greater internal than external hazard.

SAQ 1.2a	The following statements refer to the α emitting isotope $^{239}_{94}$Pu or to its α particles. Indicate whether the statements are true (T) or false (F).

(*i*) All α particles from $^{239}_{94}$Pu have the same energy.

T / F

(*ii*) α spectrometry is a possible means of identifying $^{239}_{94}$Pu.

T / F

(*iii*) The neutron : proton ratio is lower in $^{235}_{92}$U than in $^{239}_{94}$Pu.

T / F

(*iv*) α particles from $^{239}_{94}$Pu travel long distances in body tissue.

T / F

(*v*) α particles from $^{239}_{94}$Pu are *not* easily measured by Geiger counters.

T / F.

1.2.2. Negatron Emission

There are several decay processes known collectively as beta decay. The most common of these is a nuclear process for converting a neutron to a proton; in order to retain electroneutrality an electron must be produced also.

$$^{1}_{0}n \rightarrow ^{1}_{1}p + ^{0}_{-1}e$$

(The process is actually rather more complicated than this, but this does not affect the properties, which we may wish to use, that derive from the process).

The electron cannot remain in the nucleus; it is ejected at nearly the speed of light. In this context it is widely known as the negative beta particle or *negatron*, and is given the symbol β^-.

From the viewpoint of usage of β^- emitters it is again (as for α emitters) instructive to consider relevant aspects of the energetics of β^- emission.

(*a*) For reasons which are beyond this treatment, there is a *spectrum* of β^- energies for the negatrons emitted from a given isotope; the energy is most often specified by the maximum value, E_{max}, or as the mean value, E_{mean}, where E_{mean} is roughly $E_{max}/3$. A typical example is shown in Fig. 1.2a.

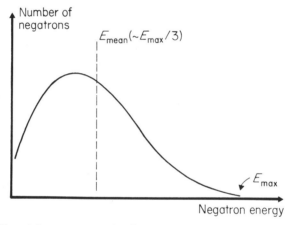

Fig. 1.2a.   *A typical negatron energy spectrum*

(*b*) The energy range for β^- particles (particularly those of the more widely used isotopes) is rather lower than that for α particles. Some typical values are shown in Fig. 1.2b.

Isotope	^3H	^{14}C	^{32}P	^{35}S	^{45}Ca	^{60}Co
$E_\beta(max)/MeV$	0.019	0.155	1.710	0.167	0.258	0.316

Fig. 1.2b.   *Some widely used negatron emitters*

(*c*) The values (Fig. 1.2b) are for isotopes for which each nucleus emits negatrons with only a single value of E_{max}. However, the situation is often more complicated; for example, for the widely used isotope ^{131}I, the predominant β^- has $E_{max} = 0.607$ MeV, but the isotope also emits smaller proportions of negatrons with $E_{max} = 0.337$ and 0.810 MeV respectively.

(*d*) Sometimes almost the entire energy associated with the conversion $^1_0n \rightarrow {}^1_1p$ is carried away by the negatron, in which case the isotope is known as a 'pure negatron emitter'. The first five isotopes in Fig. 1.2b are of this type. However, much more commonly the daughter nucleus is formed in an excited state, and reaches stability by emitting photons of energy, known as γ rays (see 1.2.5). Thus, many isotopes are *mixed β^- and γ emitters*.

Hopefully you are now realising that the overall decay scheme for many isotopes is very complex. It is not necessary to pursue this point, except to stress that a fundamentally important step in considering the possible usage of a radioactive isotope is to interpret the published decay scheme, particularly to identify the predominant ray/particle and hence work out the best detection method. You may recall also that the principal reason why α emitting isotopes are little used as tracers is that the α particles have little penetrating power. Now, although β^- particles in general have lower initial energy, they are much more penetrating because they cause far less ionisation per unit of distance travelled. Why? Because not only do they have the same charge as the electrons of surrounding atoms but they also have *far* less mass than α particles (roughly 1/7358).

As a typical example an α particle of 3 MeV initial energy will penetrate roughly 0.05 mm of aluminium, whereas a negatron of the same initial energy will penetrate roughly 6.5 mm Al.

What might we suggest as the most likely practical consequences of such data?

(*a*) Because most negatrons (except those of low initial energy) can fairly readily penetrate materials such as glass

and mica, which are used in constructing Geiger counters, it follows that we should be able to use such counters for measuring sources which emit medium and high
energy negatrons.

(*b*) Conversely we shall not use such counters for low energy
negatrons.

(*c*) Although negatrons are a greater external health hazard
than α particles they do not require especially dense
shielding.

In practice, negatron emitting isotopes find wide application as tracers in the study of analytical processes such as solvent extraction,
ion exchange, and solubility studies.

SAQ 1.2b   The following statements refer to the β^- emitting isotope $^{32}_{15}P$ or to its negatrons.

(*i*)   All negatrons from ^{32}P have the same energy.

(*ii*)   Geiger counting is not a possible means of
measuring ^{32}P.

(*iii*) The neutron : proton ratio is lower in $^{32}_{16}S$
than in $^{32}_{15}P$.

(*iv*)   Negatrons from ^{32}P are a greater external
health hazard than most α particles.

(*v*)   The product of β^- emission from $^{32}_{15}P$ is
$^{32}_{14}Si$.

Work out which *three* of these statements (*i*)–
(*v*) are *false*. $\longrightarrow$

SAQ 1.2b

1.2.3. Positron Emission

A second form of beta decay is known as positron emission. If, in order to achieve stability it is necessary to increase the n : p ratio, one possible mechanism is the direct conversion of a proton to a neutron. Electroneutrality must be maintained, thus suggesting the production of a positive electron (a positive beta particle) or *positron*.

$$_1^1\text{p} \rightarrow \,_0^1\text{n} + \,_{+1}^0\beta$$

You may find the concept of a positive electron rather surprising!

As for β^- formation, this process leading to β^+ formation is more complex than this simple equation suggests; for instance we have ignored energy changes during the process. The β^+ particle is ejected from the nucleus, loses energy by collision with other atoms, and then combines with an extranuclear electron. The total mass of $(\beta^+ + \text{e}^-)$ is converted completely to energy. Ask yourself how we can relate mass to energy; the answer is through the Einstein equation $E = mc^2$ (1.1.3). You may be surprised how easy it is to use!

∏   Calculate the energy produced by the complete conversion
     of the mass of $(\beta^+ + e^-)$.

     Hint: the trick is to use the correct SI units.

     speed of light   = 2.998×10^8 ms^{-1}

     mass of electron   = 9.1095×10^{-31} kg

     total mass of β^- + electron   = 18.219×10^{-31} kg

     $\therefore \quad E \;=\; mc^2 = 18.219 \times 10^{-31} \times (2.998 \times 10^8)^2$

     $ \;=\; 1.6375 \times 10^{-13} \text{ kg m}^2 \text{ s}^{-2} \text{ (J)}$

     $ \;=\; \dfrac{1.6375 \times 10^{-13}}{1.6022 \times 10^{-19}} \text{ eV}$

     $ \;=\; 1.022 \times 10^6 \text{ eV} = 1.022 \text{ MeV}$

The energy is liberated as two identical photons (called gamma rays)
each of 0.511 MeV. These are quite penetrating, and so experimentally it is quite easy to detect them and hence to use positron-
emitting isotopes as tracers.

1.2.4. Electron Capture (EC)

This also falls under the general heading of beta decay. You have
just calculated that for the process

$$\beta^+ + e^- \rightarrow 2\ \gamma \text{ rays}$$

the energy released is 1.022 MeV. What happens if we have an isotope which is rich in protons, but for which the energy change,
proton → neutron, is *less than* 1.022 MeV? Clearly we cannot have
β^+ emission; an alternative is

$$^1_1\text{p} + \,^{\ 0}_{-1}\text{e} \rightarrow \,^1_0\text{n}$$

Of course this requires the nucleus to capture an electron, hence the title. Perhaps you might predict that the electron will be from an inner orbital. In your earlier studies you may have seen such orbitals described as being in the 'K shell', and for this reason electron capture is also known as K-capture. When the electron is captured by the nucleus there is technically a 'vacancy' in the orbital concerned, but this is immediately filled by an electron dropping down from the orbital of the next highest energy, and so on until the 'vacancy' is in the outermost orbital. (Of course there isn't really a vacancy because in the process $_1^1p + _{-1}^0e \rightarrow _0^1n$ the atomic number has decreased by one). The energy which is liberated when the electrons rearrange lies in the X-ray region, so what is actually measured for an electron capture isotope is the emission of X rays.

Once again, EC isotopes can be quite useful; one good example which is particularly widely used in radioimmunoassay (see 4.3) is ^{125}I, although this also emits low energy gamma rays.

1.2.5. Gamma Emission

You have already seen that the emission of alpha or beta particles may leave the daughter nucleus in an excited state, and that the nucleus will then lose energy equivalent to the difference between the energy of the nuclear ground state and the excited state.

This short wavelength radiation has already been called gamma radiation.

Before we consider practical aspects of using γ ray emitting isotopes there are two theoretical points that need comment.

Firstly, the simple treatment in (1.2.2d) inferred the emission of a single β^- particle and a single γ ray, and for some isotopes this is the case. Thus, $^{198}_{79}Au$ emits a β^- of $E_{max} = 0.961$ MeV leading to an excited state of $^{198}_{80}Hg$ which then emits a γ ray of $E = 0.412$ MeV leading to the ground state of $^{198}_{80}Hg$. *But*, more than one β^- (or α) particles of different energies may be emitted, thus leading to more than one excited state of the daughter isotope. This will necessarily lead to the emission of more than one γ ray *of differing energy* as the ground state is reached.

For example, the decay scheme for $^{130}_{53}$I is given in tabular form as:

β^-/MeV 0.60 (54%), 1.01 (46%)

γ/MeV 0.41 (24%), 0.53 (100%), 0.66 (100%), 0.74 (70%)
1.15 (30%)

The percentages in brackets represent the relative number of emissions of that particular type and energy.

The decay scheme becomes clearer when it is drawn diagrammatically (Fig. 1.2c)

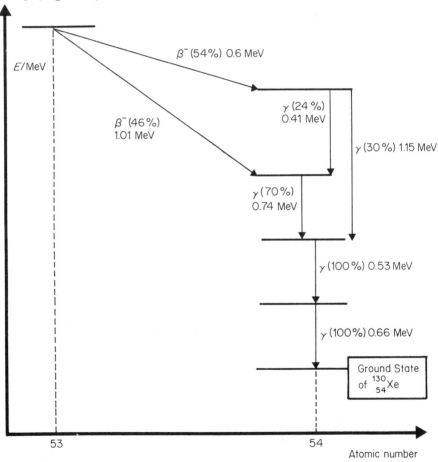

Fig. 1.2c. *Decay scheme for $^{130}_{53}$I (not to scale)*

In this case there are *four* excited states of $^{130}_{54}\text{Xe}$. The total energy loss on β^- and γ emission from $^{130}_{53}\text{I}$ is $(0.60 + 0.41 + 0.74 + 0.53 + 0.66) = 2.94$ MeV.

We shall see later that we can make good practical use of the fact that isotopes emit γ *rays of characteristic energy* as shown in Fig. 1.2c, rather than as a continuous spectrum.

Perhaps the second point of theory is more obvious. Since γ rays have no mass they would be expected to be much more penetrating than (say) β^- particles of similar energy, and this is indeed true.

What are the practical consequences of these points?

(*a*) If we can set up counting apparatus which allows us to discriminate between γ rays of different energy we should be able not only to identify the presence of individual gamma emitters *but to count them independently in a mixture*. This technique, known as γ ray spectrometry, is of fundamental importance in radiochemical analysis.

(*b*) Because γ rays (particularly of medium and high energy) are very penetrating they can be quite a significant health hazard, and γ emitting isotopes normally require quite thick shielding; lead is widely used.

(*c*) For similar reasons to (*b*), γ ray emitting isotopes are not usually measured by Geiger counting; the γ ray penetrates completely through the counter without causing sufficient ionisation of the filling gas (the process on which such a counter depends (3.1)). Nevertheless, this will not inhibit the Geiger counter from detecting any associated β^- emission.

Putting these points together, we might conclude that whilst isotopes emitting both β^- particles and γ rays have some disadvantages (particularly in relation to hazard) relative to pure β^- emitters, they are nevertheless likely to find widespread applications as tracers in analytical processes.

SAQ 1.2c

The decay scheme for $^{38}_{17}$Cl may be represented as:

β^-/MeV : 1.11 (31%), 2.71 (16%), 4.81 (53%)
γ /MeV : 1.60 (31%), 2.10 (47%)

Draw this scheme diagrammatically, plotting energy on a vertical scale and atomic number on a horizontal scale, and hence identify the daughter product.

1.2.6. Other Types of Decay

Isotopes of naturally occuring elements do not *spontaneously* undergo other forms of decay. Some isotopes of the synthetic transuranic elements are found to undergo spontaneous fission (ie the nucleus splits into two parts). A good example is $^{252}_{98}Cf$ (Cf = californium); neutrons are released upon fission of this isotope, and in recent years it has achieved widespread use as a source of neutrons for activation analysis. It is discussed in more detail in Section 4.4.

Nuclear reactions can, of course, be *induced* in several ways. The most familiar are nuclear fission of, for example, $^{235}_{92}U$,

$$^{235}_{92}U + {}^1_0n \rightarrow \text{Fission Products} + \text{av } 2.5{}^1_0n + \text{Energy}$$

and neutron emission,

$$^9_4Be + {}^4_2\alpha \rightarrow {}^{12}_6C + {}^1_0n$$

Both of these reactions find preparative and analytical use and are discussed later.

SUMMARY AND OBJECTIVES

Summary

Nuclei undergo radioactive decay in order to attain a more stable neutron : proton ratio, and to lower the overall energy of the nucleus. Although α particle emission is important for the heavier elements the more widespread types of decay are the various forms of β decay, and γ ray emission; it is these latter types of decay that are of the greatest analytical importance, principally because properties of the emitted particles and/or rays lend themselves to rather easy measurement.

Objectives

You should now be able to:

- specify those properties of practical importance of the particles/rays emitted from nuclei;

- explain why individual properties are important;

- appreciate the significance of the differences in properties of various particles/rays;

- relate the properties to practical applications such as detection and measurement.

1.3.   LAWS OF RADIOACTIVE DECAY

Overview

This section is intended to give a simple derivation of the fundamental decay law for the rate of decay of a single radioisotope. Following from this law there is a need to relate rate of decay to a simple parameter—the 'half-life'. A preliminary consideration is given to the significance of half-life values.

The mathematical treatment of radioactive decay that is most often important in analytical chemistry is concerned with the simple decay of a single radioisotope, and hence this section concentrates on such individual decay. More complicated decay schemes such as those which occur in a natural decay series or in a fission product decay chain are rarely of analytical significance. The most frequently occurring situation in which the analyst is likely to meet multi-isotopic mixtures is in activation analysis (Section 4.4). In this case the activities can normally be measured independently, and the simple theory of individual decay can be used.

1.3.1.   Radioactive Decay Laws

It is perhaps easiest to say what the rate of decay does *not* depend on! Let us suppose that we have a sample containing the negatron emitter, ^{14}C. The state of chemical combination will have no effect on the rate of decay; this will be the same for ^{14}C in $^{14}CO_2$, $^{14}CH_4$, $^{14}C_6H_6$ or the most complicated organic molecule. Similarly, physical conditions such as temperature and pressure have no effect, nor does the allotropic state (graphite, diamond etc). Individual ^{14}C nuclei do not transmit their activity to other nuclei.

Thus we may conclude that in any array of identical radioactive nuclei the disintegration of any individual nucleus is completely independent of the other nuclei. Hence it is necessary to treat this *random* decay by simple statistical methods.

IF YOUR MATHEMATICAL BACKGROUND IS NOT ADE-
QUATE FOR THE TREATMENT WHICH FOLLOWS, DON'T
WORRY—IT IS THE RESULT WE OBTAIN THAT IS IMPOR-
TANT.

The first step is to assign to every nucleus in an array of identical
radioactive nuclei the same *probability* of disintegrating in a given
time. This probability is given the Greek symbol 'lambda' (λ), and
is called the *decay constant.*

The next step is to assume that if at time t there are N radioactive
nuclei in the array then in a small interval of time dt, the number
that will have decayed is dN. Thus, we have:

$$dN = -\lambda N \, dt$$

which may be written as:

$$dN/dt = -\lambda N \qquad (1.3a)$$

This differential equation which you may recognise as a typical first
order kinetic equation is often known as the *fundamental decay
equation*; note the minus sign, which occurs since the number of
radioactive nuclei is decreasing. The term dN/dt is often called the
activity of the sample.

Finally we can integrate the differential equation (1.3a), putting N
$= N_0$ when $t = 0$, and this gives:

$$N = N_0 \, e^{-\lambda t} \qquad (1.3b)$$

It is not normally convenient to measure the absolute numbers of
radioactive nuclei. Experimentally it is the activity of a sample which
is measured, and since the activity (given the symbol A) is directly
proportional to N, the above equation (1.3b) becomes:

$$A = A_0 \, e^{-\lambda t} \qquad (1.3c)$$

(In section 1.4 we shall discuss units of A, but for the present we note that the simplest unit is disintegrations per second). There are various ways of plotting this *exponential* decay curve. A simple plot of A/A_0 against t is shown in Fig. 1.3a.

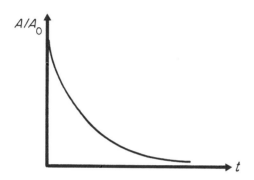

Fig. 1.3a. *Exponential decay of activity*

However, most people (not least students!) are happier when dealing with straight line graphs. Hence, it is much more common (and as it happens much more useful) to plot the logarithm of A (either to base 10, or the natural logarithm) against t. Taking logarithms of Eq. 1.3c:

$$\ln A \;=\; \ln A_0 - \lambda t \tag{1.3d}$$

$$\text{or} \quad 2.303 \log A \;=\; 2.303 \log A_0 - \lambda t \tag{1.3e}$$

Since the equations are of the form $y = mx + c$ we shall expect a plot of $\ln A$ (ie y) against t (ie x) to be a straight line as in Fig. 1.3b

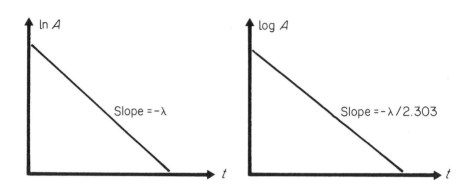

Fig. 1.3b.   *Logarithmic plots of activity versus time*

We now need to think a little about the significance of the value of λ and of its units. It may be obvious to you what the units must be, but in any case you can confirm the units by carrying out SAQ 1.3a.

SAQ 1.3a	The activity of a sample of a radioactive isotope was found to be 5000 disintegrations per second when initially measured, and 90 seconds later was found to be 1500 disintegrations per second. What are the value and the units of the decay constant?

There are two important aspects of the result you have obtained. We said earlier that a radioactive isotope can be characterised by a certain value of λ, and you have calculated just such a value for the isotope concerned. Since each radioactive isotope has a unique decay constant we might expect to find tables of radioisotopes defined according to their decay constants, in a similar way to defining (say) an oxidising agent by its standard reduction potential. In practice this is never done, and the reason leads to the second aspect we need to consider; the units you obtained for λ are s^{-1}. This is just a particular example of the general situation that, depending on what units we use for activity, λ *will have units of reciprocal time.* These are very inconvenient units, and thus a better system must be sought.

1.3.2. Half-Life

It is much simpler to characterise the decay of a radioisotope by a term with units of time rather that of reciprocal time, and this done by the concept of 'half-life'. The half-life, which is given the symbol $t_{0.5}$, is defined as the time taken for the activity of the sample to be reduced to half its initial value ie for A_0 to be reduced to $A_0/2$, for $A_0/2$ to be reduced to $A_0/4$ etc.

∏ Try the simple example of calculating the number of radioactive atoms remaining after 40 seconds for an isotope of $t_{0.5}$ = 10 s, where the initial number of radioactive atoms was 1600.

Your answer should be 100; after 10 s, N = 800; after 20 s, N = 400; after 30 s, N = 200; and after 40 s, N = 100.

Let us now do some further arithmetic:

If we put $A = A_0/2$ and $t = t_{0.5}$ into (1.3e):

$$A_0/2 \quad = \quad A_0 \exp(-\lambda\, t_{0.5})$$

$$\therefore \quad \exp(\lambda t_{0.5}) \quad = \quad 2$$

$$\therefore \quad \lambda \, t_{0.5} \quad = \quad \ln 2$$

$$\therefore \quad t_{0.5} \quad = \quad 0.693/\lambda \qquad (1.3f)$$

We can now see the value of the linear plot of log A against t; it enables us to calculate λ from the slope of the graph, and hence to calculate $t_{0.5}$.

You may now be able to appreciate that it is far simpler to characterise an isotope by its value of $t_{0.5}$ rather than of λ. Thus, for example, for $^{32}_{15}P$ we can say that $t_{0.5} = 14.3$ days, and have a much better 'feel' for what this means than if we say that the decay constant for $^{32}_{15}P$ is 0.0485 days^{-1}.

SAQ 1.3b

The data below refer to the decay of a sample of ^{234}Pa.

A/disintegrations s^{-1}	237	184	160	129	111	91	77	63
t/s	0	26	42	62	80	100	116	138

By drawing a suitable graph of these data calculate:

(i) the decay constant and

(ii) the half-life for this isotope.

SAQ 1.3c	The half-life of the isotope $^{24}_{11}$Na is 15.00 h. The disintegration rate of a sample of this isotope was 16 000 disintegrations min^{-1} at 09.00 h on a certain day. At 21.00 h on the following day the disintegration rate (disintegrations min^{-1}) was:

    (*i*) 9191   (*ii*) 13548   (*iii*) 3032.

We shall see later in this unit, and particularly in the sections concerned with the practical aspects of usage and disposal of radioisotopes, how important it is to be able to interpret half-life values. You should also bear in mind that we can develop a relationship between $t_{0.5}$ and mass, and this is described in Section 1.4 on units of activity.

1.3.3. More Complex Decay

Although they are not treated here you should recognise that there are several situations in which the decay of a radioisotope cannot be considered quite as simply as the foregoing parts suggest. Amongst these are:

Successive decay—in which a parent decays to a daughter which itself decays further.

Branching decay—in which the parent decays by different routes to two (or more) daughter products.

Mixed decay—the decay of a mixture of non-related radioisotopes.

1.3.4. Rate of Emission of Individual Particles/Rays

Although it is not of direct analytical importance we should note that in 1.3.1 and 1.3.2 we were concerned with the gamma rays. Thus if the disintegration of a nucleus results in the emission of one β^- particle and two gamma rays then a disintegration rate of 10^3 disintegrations per second by the nucleus will result in the emission of 10^3 negatrons per second *and* 2×10^3 gamma rays per second. This will be very important in calculations of dose rate and of the nature and thickness of shielding materials.

∏   Can you remember why β^- decay is often followed by γ radiation?

Because quite often the β^- particle does not carry away all the emitted energy. If you could not remember check back to 1.2.2d.

SUMMARY AND OBJECTIVES

Summary

The rate of decay of a radioisotope is a random event and is treated statistically. Each radioisotope has a unique probability of decay known as the decay constant. The mathematical treatment of radioactive decay leads to an exponential decay law. The most convenient parameter for expressing the rate of decay of an individual radioisotope is the half-life.

Objectives

You should now be able to:

- explain the terms involved in the fundamental decay law, and their relationship to each other;

- define the term half-life;

- relate the half-life to the decay rate and decay constant;

- calculate values of half-life from experimental data;

- understand the problems involved in quantitatively explaining more complex decay schemes.

1.4.   ACTIVITY—UNITS AND DEFINITIONS

Overview

This section is intended to introduce you to those units of activity in current usage, and the relationship between them. The need to relate mass to activity for both pure and diluted samples of radioisotopes is explained, and quantitative aspects are presented.

In Section 1.3, you have seen that the rate of radioactive decay is known as the 'activity' of the sample. It is now necessary to consider the various ways of expressing the activity, and to consider how it may be related to the weight or chemical concentration of the sample.

1.4.1.   Units of Activity

The simple unit of activity on the SI system is the becquerel (Bq), where 1 Bq is 1 disintegration per second. Unfortunately the situation is more complicated than this, partly because in a practical situation one becquerel is an inconveniently *small* quantity (a typical tracer experiment might well involve disintegration rates of 10^3 Bq), but also because there is still a strong tradition of using the pre-SI unit, the curie (Ci) where 1 Ci is 3.70×10^{10} disintegrations per second. Some simple arithmetic will tell you that for tracer experiments the curie is an inconveniently large quantity. (The historical background to the curie is given in 1.4.2).

∏   To see if you fully understand the units, convert 10^3 disintegration s^{-1} (ie 10^3 Bq) to curies.

1 Bq $\equiv$ 1 disintegration s^{-1}

$$\equiv \frac{1}{3.7 \times 10^{10}} \text{ Ci} \equiv 2.70 \times 10^{-11} \text{ Ci};$$

thus 10^3 Bq $= 2.7 \times 10^{-8}$ Ci.

There are two important practical consequences of the use of two systems of units:

(*a*) It is common practice in textbooks, journals, and catalogues to quote activities in both becquerels and curies.

(*b*) It is frequently necessary to use prefixes to the main unit in order to conveniently express the activity. These prefixes are the ones which are used for any SI unit—not just units of radioactivity. Those in common use in radioactive work are tabulated:

Factor	Prefix	Symbol
10^{12}	tera	T
10^{9}	giga	G
10^{6}	mega	M
10^{3}	kilo	k
10^{-3}	milli	m
10^{-6}	micro	μ
10^{-9}	nano	n
10^{-12}	pico	p
10^{-15}	femto	f

Given the small value of 1 Bq it may not surprise you that in analytical usage particularly it is often necessary to use kBq and MBq, and occasionally GBq. Correspondingly the relevant suffixes are often used before the curie, mCi and μCi being particularly useful.

Although it is likely that the becquerel-based system will eventually become standard it is necessary for you to be familiar with the interconversions.

Some of the more common conversions are tabulated:

1 Ci	1 mCi	1 μCi
3.7×10^{10} Bq	3.7×10^{7} Bq	3.7×10^{4} Bq

1 Bq	1 kBq	1 MBq
2.7×10^{-11} Ci	2.7×10^{-8} Ci	2.7×10^{-5} Ci

A further slight complication is that experimentally it is very common practice to measure activities as 'counts' per minute (in fact allowing for the counter not being 100% efficient).

SAQ 1.4a	A sample gave an experimental count rate of 9800 counts minute^{-1} in a counter of 30% efficiency. Choose from options (i)–(iv) the activity expressed as Bq and Ci respectively:
	(i) 1.96 MBq or 52.97 μCi
	(ii) 544.44 Bq or 14.71 nCi
	(iii) 32.67 kBq or 0.88 μCi
	(iv) 9.80 kBq or 0.26 μCi

1.4.2. Activity–Mass Relationships

You should ask yourself why the unit of activity that has been so widely used for many years was chosen—what is the point of a unit defined by 1 Ci = 3.7×10^{10} disintegrations s^{-1}? As you may guess the name of the unit is the surname of two of the most prominent early workers on radioactivity, Pierre and Marie Curie. Amongst their many discoveries Marie Curie in 1902 isolated 0.1 g of pure radium chloride from over one tonne of uranium ore waste. This amazing achievement was a conclusive proof of the existence of the then new element, radium.

Historically the curie was defined as the number of decays per second from one gram of ^{226}Ra. This value assumed that the half-life of ^{226}Ra is 1580 years, and this is now known not to be precisely correct. The detail of this argument is not directly important; what is important is that it suggests a relationship between the *mass of a radioisotope and its activity.*

Let us return to Eq. 1.3a:

$$dN/dt = -\lambda N \qquad (1.3a)$$

From the Avogadro constant ($N_A = 6.022 \times 10^{23}$ mol^{-1}) we should be able to calculate the number of atoms N in any sample of a *pure* radioisotope. We also know the value of the decay constant, Hence we should be able to derive a relationship between dN/dt (ie activity) and the weight of the sample.

Let us assume we have a 1 gram sample of *pure* $^{32}_{15}$P and that the relative atomic mass of this isotope is 32.00

$$\text{Then } 1\text{ g} \quad = \quad 1/32 \text{ moles of } {}^{32}_{15}\text{P}$$

$$= \quad \frac{1 \times 6.022 \times 10^{23}}{32} \text{ atoms of } {}^{32}_{15}\text{P}$$

$$\text{Now } t_{0.5} \quad = \quad 14.3 \text{ days}$$

$$= \quad 14.3 \times 24 \times 60 \times 60 \text{ s}$$

$$\therefore \quad \lambda \quad = \quad 0.693/14.3 \times 24 \times 60 \times 60 \text{ s}^{-1}$$

$$\therefore dN/dt \quad = \quad (-)\frac{0.693 \times 6.022 \times 10^{23}}{32 \times 14.3 \times 24 \times 60 \times 60} \text{ disintegrations s}^{-1}$$

$$= \quad 1.0555 \times 10^{16} \text{ disintegrations s}^{-1}.$$

ie the activity of 1 g of pure $^{32}_{15}$P

$$= \quad 1.0555 \times 10^{16} \text{ Bq } (\equiv 2.853 \times 10^5 \text{ Ci})$$

We can do identical calculations for any pure radioisotope, and for any pure compound of that isotope, and some values of the activity of 1 g of certain widely used radioisotopes are tabulated.

Isotope	Activity/Bq
^3H	3.57×10^{14}
^{14}C	1.65×10^{10}
^{35}S	1.58×10^{15}
^{90}Sr	5.16×10^{12}
^{238}U	1.24×10^4

SAQ 1.4b

The radioisotope ^{125}I for which $t_{0.5} = 60$ days is widely used in analytical procedures.

Calculate

(*i*) the activity (in Bq) of 1 g of pure ^{125}I

(*ii*) the mass of a sample of ^{125}I with an activity 10^3 Bq.

What is the significance of these data, and particularly of the answer to the second part of SAQ 1.4b?

> *For all radioisotopes except those with extraordinarily long $t_{0.5}$ values a minute mass gives a very high activity.*

Put another way, the amount of activity needed for a typical tracer experiment (10^3 Bq) is associated with an almost insignificant mass—you have just calculated that the activity of 1.55×10^{-12} g of ^{125}I is 10^3 Bq. Thus if we can measure 10^3 Bq accurately we have the potential for highly sensitive analysis.

1.4.3. Carriers and Specific Activity

Your general knowledge of analytical chemistry should tell you that the experimental difficulties of working with quantities such as 10^{-12} g are very great, and it is therefore not surprising that in most radiochemical experiments the active isotope (normally as a chemical compound) is mixed with inactive materials (normally, though not always, inactive isotopes of the same element in the same chemical form as the active isotope).

Such materials are called *Carriers*; a sample of a pure radioisotope undiluted with any other material is called *Carrier-Free*.

Note carefully that this suggests that the carrier is deliberately added to help chemical manipulation, and indeed this normally is the case. But you should also be clear that in the nature of the preparation of many radioisotopes it is very difficult to obtain them carrier-free initially—(see Section 2.1).

You may now appreciate that normally it is necessary to define not just the radioactive concentration in isolation, but its value relative to the total chemical concentration of radioisotope plus carrier. This is known as the *Specific Activity*.

In the same way that units of activity are in transition from the curie to the becquerel, so there are problems for units of specific activity because of the different ways of expressing chemical concentrations.

The preferred SI units are Bq mol^{-1} or Bq g^{-1} for solids, or Bq dm^{-3} for solutions. However, much current literature still uses Ci mol^{-1}, Ci g^{-1}, Ci dm^{-3} or some sub-unit of these such as mCi mmol^{-1}.

SAQ 1.4c	A sample of the amino acid glycine

SAQ 1.4c

A sample of the amino acid glycine

$$H_2NCH_2COOH,$$

labelled with ^{14}C, is available at a specific activity of 20 mCi mmol^{-1}. Which of the options (*i*)–(*iii*) is the correct value for the specific activity:

(*i*)   9.86 MBq g^{-1};

(*ii*)   9.86 TBq g^{-1};

(*iii*) 9.86 GBq g^{-1}?

SUMMARY AND OBJECTIVES

Summary

The activity of a radioactive isotope is most fundamentally expressed in units of disintegration per second, and on the SI system this is given the name 'one becquerel'. The pre-SI unit is the curie, where one curie is 3.7×10^{10} disintegrations per second. For virtually all radioactive isotopes very small mass is associated with very high activity. Hence for experimental purposes inactive material is mixed with the active isotope, and it is then necessary to define activity in relation to chemical concentration—the specific activity.

Objectives

You should now be able to:

- specify and interconvert the currently accepted units of activity;

- understand the need to use fractions or multiples of these units in practical situations;

- explain the relationship between mass and activity for a pure radioisotope;

- recognise the need for different definitions of activity for radioisotopes diluted with stable isotopes;

- perform calculations involving the various units of activity.

2. Radioisotopes and Labelled Compounds

2.1 THE PREPARATION AND AVAILABLITY OF RADIOACTIVE MATERIALS

Overview

This section surveys the availability of radioactive materials, both natural and synthetic. It describes routes to synthetic radioisotopes, summarises the forms in which such isotopes are commercially available, and relates the nature of the product to potential use.

2.1.1. The Preparation of Radioactive Materials

We saw in 1.2.1. that some elements for which *all* the isotopes are radioactive occur naturally; these are the heavier elements such as uranium. There are other elements for which only one of the naturally occurring isotopes is radioactive. A good example here is potassium, which contains 0.012% of radioactive ^{40}K; since potassium is an essential constituent of the body it follows that we all contain some ^{40}K, which for normal human beings is maintained at an equilibrium level of roughly 0.1 μCi (3.7 kBq).

If we need a sample of a naturally occurring radioactive element we are limited only by our ability as chemists to separate it from the materials with which it occurs, and to purify it to the required level. We mentioned in 1.4.2 the historic separation of radium; a modern example is the processing of uranium for nuclear reactor fuel, from an ore content quite possibly as low as 0.1% U to an eventual purity of at least 99.99%.

However, the overwhelming majority of radioactive isotopes with which analytical chemists are likely to come into contact are man-made, and so there is a need for their preparation from stable materials, and their subsequent chemical conversion to the required form.

With relatively few exceptions the materials concerned are available commercially. In the UK the principal supplier is Amersham International plc (formerly the Radiochemical Centre Ltd., and at one time a subsidiary of the UK Atomic Energy Authority), but other organisations also offer a service (eg ICI plc, certain Universities).

The preparation and processing of radioactive materials can be very complex, and since in this course we are principally concerned with the analytical *uses* of radioactive isotopes we shall only outline the topic.

The Irradiation of Stable Materials

You have seen in the earlier sections that nuclei are radioactive because they are moving from a state of high energy to one of lower energy, normally to achieve a more favourable balance of neutrons and protons. So how would you guess that we might think of *making* radioisotopes? In principle the answer is simple: put in energy, and at the same time produce an unbalanced neutron : proton ratio.

In practice this is most commonly done by irradiating a stable target with neutrons, on a commercial scale in a nuclear reactor. We shall consider this topic in more detail in Section 4.4 in its analytical context. All we need say here is that neutron activation is particularly suitable for a wide range of elements, and—because of the high neutron fluxes that can be obtained in a reactor (10^{12}–10^{18} neutrons s^{-1} m^{-2})—quite high levels of activity can be obtained.

So what are the practical requirements? The first is the *size* of the sample. A typical size (often called a 'standard can' because of the nature of the container) is a cylinder about 7 cm in length by 2.25 cm internal diameter. This is not as restrictive as it may sound. Such a container could accommodate many grams of most solid materials, and as an example 0.1 g of gold irradiated at 10^{16} neutrons $s^{-1}m^{-2}$ for 1 week would give approximately 600 mCi of radioactive ^{198}Au: enough for a million tracer experiments!

The second restriction is the *physical nature* of the material. By far the preferred materials are solids that are stable to heat and not susceptible to radiation damage. Heat is produced in the fission reaction, and depending on the design of the reactor, this can be a problem during irradiation. An important side effect of neutron bombardment is the breaking of bonds, and care is necessary that volatile products are not produced. Under rather restrictive circumstances it is possible to irradiate liquids, but as you may expect, it is very unusual to irradiate gases—their low density is not compatible with a small volume, and thermal expansion is a problem.

Chemical restrictions also are quite severe; perhaps you might predict that target materials should be as chemically inert as possible. Ionic solids are preferred to covalent compounds, because the side effects of radiation damage are minimised in ionic lattices. Again you might predict that the simpler the chemical form of the target the better.

There is one further problem which is directly concerned with the nature of the activation process. When an inactive nucleus 'captures' a neutron more than one product nucleus may be formed. As an example consider the bombardment of chlorine nuclei with neutrons.

$$^{35}_{17}Cl + {}^{1}_{0}n \rightarrow {}^{36}_{17}Cl + \gamma$$

$$^{35}_{17}Cl + {}^{1}_{0}n \rightarrow {}^{35}_{16}S + {}^{1}_{1}p$$

$$^{35}_{17}Cl + {}^{1}_{0}n \rightarrow {}^{32}_{15}P + {}^{4}_{2}\alpha$$

All these products are radioactive, and chemical separations will be necessary if a chloride is irradiated. To minimise this problem it is usual to irradiate *either* the free element or the element combined with another element that is not easily activated (O,N,C) eg the oxide, nitride, or carbonate.

SAQ 2.1a

> Consideration is being given to the irradiation in a nuclear reactor of each of the following elements. For each element, place a tick in the appropriate column relating to the feasibility of irradiation.
>
> | | Definitely Yes | Definitely No | Possibly |
> |---|---|---|---|
> | Fluorine | | | |
> | Sodium | | | |
> | Aluminium | | | |
> | Phosphorus | | | |
> | Iodine | | | |
> | Gold | | | |

SAQ 2.1b	From SAQ 2.1a you should have appreciated that direct reactor irradiation of sodium metal is unfavoured. From the list of compounds which follow select the most appropriate target materials for preparing radioactive sodium and give your reasons:
	NaCl; NaOH; Na_2O_2; Na_2CO_3; Na_3PO_4.

Finally, let us consider how we might prepare isotopes which have a neutron : proton ratio that is *lower* than that of the stable isotope(s) of that element; some of these isotopes are very useful as tracers, so they are worth preparing!

∏   What are the most likely ways in which such isotopes will decay? If you can't remember, or work it out, check back to 1.2.3 and 1.2.4.

Hopefully it is clear that if we wish to *lower* the neutron : proton ratio we are unlikely to be able to do it by neutron bombardment. Normally the desired isotopes are produced by bombarding a suitable target with (usually quite simple) charged particles which are raised to the appropriate energy in some form of accelerating machine. The most widely used version accelerates the particles in a spiral path, and is known as a 'cyclotron'.

∏ To illustrate some typical cyclotron reactions, complete the following equations:

$$^{24}_{12}Mg + {}^{2}_{1}H \rightarrow {}^{22}_{11}Na +$$

$$^{56}_{26}Fe + {}^{2}_{1}H \rightarrow \qquad + {}^{4}_{2}He$$

$$+ {}^{2}_{1}H \rightarrow {}^{75}_{34}Se + 2\,{}^{1}_{0}n$$

The answers are $^{4}_{2}He$, $^{54}_{25}Mn$, and $^{74}_{33}As$ respectively. What is the most obvious point about the main product of each reaction? It is an isotope of a *different element* from the target. Thus it should be possible to separate it chemically from the target with the potential bonus of getting the product with *a high specific activity*. As a further bonus it sometimes happens that the product has a more useful half-life than other isotopes of the element; good examples are:

$$^{54}Mn\ (t_{0.5}\ 312\ days)\ \text{relative to}\ ^{56}Mn\ (t_{0.5}\ 2.58\ hours)$$

and

$$^{57}Co\ (t_{0.5}\ 270\ days)\ \text{relative to}\ ^{60}Co\ (t_{0.5}\ 5.27\ years).$$

Separation of Fission Products

The fission of $^{235}_{92}U$ produces a very extensive range of radioactive fission products, which in principle are very useful for tracer work. However, their separation would require quite difficult processing of the highly radioactive spent fuel from reactors, and with a very few exceptions this is not normally a viable option.

2.1.2. Processing of Active Material

You have seen in 2.1.1 that the activated target is likely to be either a (metallic) element or a simple compound such as an oxide, nitride, or carbonate. Sometimes the material is made available to the user in this form, but more commonly it is processed to a range of different compounds or sources (see 2.1.3).

The most important criteria for the reactions used are:

(*a*) the conservation of a high specific activity, and

(*b*) the specificity of position of labelling within the molecule.

It is not possible to give details here, except to say that chosen reactions should preferably be of high chemical yield, be rapid, have few or no undesirable side products, and be capable of stringent process control. Demanding quality control of the product will be necessary also.

Important areas of development include the labelling with ^{14}C and ^{3}H of organic compounds of increasing complexity, and, partly stimulated by radioimmunoassay (Section 4.3)—labelling with halogen radioisotopes.

For ^{14}C the activation reaction is:

$$^{14}_{7}N + {}^{1}_{0}n \rightarrow {}^{14}_{6}C + {}^{1}_{1}p$$

You should recall that the target material should preferably be a stable solid, and the compound of choice usually is AlN (aluminium nitride). After activation the ^{14}C in the target is converted to CO_2; subsequent synthetic routes must be systematic, and the early stages generally follow one of a relatively limited number of standard routes. The later stages are of course determined by the particular compound to be synthesised.

A fairly simple sequence to form different amino acids is:

$$^{14}CO_2$$

$$\downarrow$$

$$CH_3\,{}^{14}COOH$$

$$Br\,CH_2\,{}^{14}COOH$$

$$NH_2\,CH_2\,{}^{14}COOH$$
glycine

$$\begin{array}{c}{}^{14}COOH\\|\\CH_2\\|\\CHNH_2\\|\\COOH\end{array}$$
aspartic acid

$$CN\,CH_2\,{}^{14}COOH$$

$$NH_2\,CH_2\,CH_2\,{}^{14}COOH$$
β - alanine

$$CH_3\,{}^{14}CH_2OH$$

$$CH_3\,{}^{14}CH_2\,COOH$$

$$CH_3\,{}^{14}CH(NH_2)COOH$$
α - alanine

The nomenclature for labelled compounds is rather more complicated than for the unlabelled counterparts since it is usually necessary to specify the position of labelling. In the above example we would have:

glycine	= 2-amino-[1-^{14}C]ethanoic acid
α-Alanine	= 2-amino-[2-^{14}C]propanoic acid
β-Alanine	= 3-amino-[1-^{14}C]propanoic acid
aspartic acid	= 2-amino-[4-^{14}C]butanedioic acid

The position of labelling may also be important, particularly if the compound is to be used for mechanistic studies. A good simple example is:

$$^{14}CH_3OH \xrightarrow{PI_3} {}^{14}CH_3I \xrightarrow{KCN} {}^{14}CH_3CN \xrightarrow{H_2O} {}^{14}CH_3COOH$$

$$Ba\,{}^{14}CO_3 \xrightarrow{HCl} {}^{14}CO_2 \begin{array}{c}\xrightarrow{\substack{H_2\\\text{catalyst}}}\\[2em]\xrightarrow{CH_3MgBr}\end{array}$$

$$CH_3\,{}^{14}COOH$$

$$Ba\,{}^{14}CO_3 \xrightarrow{Mg/heat} Ba\,{}^{14}C_2 \xrightarrow{H_2O} {}^{14}C_2H_2 \xrightarrow[\text{catalyst}]{H_2O} {}^{14}CH_3\,{}^{14}CHO \xrightarrow{[O]} {}^{14}CH_3\,{}^{14}COOH$$

How might you confirm the position(s) of the ^{14}C label in each of the above acetic (ethanoic) acid molecules?

The simplest way would be to react a known quantity of each acid in such a way that all the CO_2 was released and measured for its radioactivity. Then $^{14}CH_3COOH$ would give inactive CO_2, $CH_3^{14}COOH$ would give 100% $^{14}CO_2$, and the $^{14}CO_2$ from $^{14}CH_3^{14}COOH$ would be 50% of the activity in the initial sample.

2.1.3. Types of Commercially Available Radioactive Source

It is unrealistic to give a detailed survey of all commercially available types of radioactive material. If as an analyst (or indeed in any area of work) you become involved in radioactive work you will probably find that the radioactive sources/chemicals you need are quite limited. We shall therefore consider only a brief survey of the commercially available materials.

One simplifying factor is to distinguish between closed (sealed) sources and open sources. A closed source is one which is carefully and permanently sealed and is used principally as an external source of radiation. An open source is one in which the radioactive material (normally a liquid or a solid) is packaged in such a way that it can be opened and mixed with the system being studied. As an example in chemistry we could distinguish between a sealed source of ^{60}Co γ rays used for cross-linking chains in polymeric materials, and an open source of ^{60}Co as (say) a solution of cobalt(II) chloride, for assessing the efficiency of separation processes involving Co(II).

Let us pursue this distinction a little further. In the first case we are using the γ radiation in our application; the fact that it derives from ^{60}Co is, in a sense, irrelevant. Put another way, we will select a sealed source for the *type* and the *energy* of the radiation it supplies. For instance, if we are interested in measuring the thickness of thin sheets of plastic it would be pointless to use highly penetrating gamma rays; a source of negatrons of the appropriate energy would be quite adequate.

In the second case we are quite specifically looking at reactions involving cobalt, and so we *do* need a source containing a radioactive isotope of cobalt. In this case we will select our isotope/chemical with such criteria as half-life and ease of counting in mind.

As a generalisation we can say that in analytical applications the greatest use is made of open sources, though we should recognise one very important exception—activation analysis, in which use is made of external sources of radiation (commonly neutrons) to induce activity in a sample (Section 4.4).

Sealed Sources

These can be broadly classified against the following criteria, or combination of them.

Type of radiation: α, β, γ, neutron, X ray.

Source strength: Specified in curies/becquerels. Can vary from μCi to MCi.

Half-life: Full range available.

Physical form: Foil, disc, cylinder etc of specified dimensions.

Possible or actual usage: Impossible to generalise. Can be designed for most applications.

Health hazard: Dose rate at the surface is normally specified.

Such sources are subject to very stringent safety regulations (eg for leakage), and full specifications of the tests are published.

Two types of sealed source which are of particular and increasing use in analytical chemistry, but which do not readily fit into the above categories are:

Mossbauer sources: low energy γ sources for Mossbauer
                     Spectroscopy

^{252}Cf sources: effectively a spontaneous fission isotope used
                     as a source of neutrons for activation analysis
                     (see Section 4.4).

∏   The specification of a low energy β^- source includes the
     following data: activity = 1 mCi, $t_{0.5}$ = 100 years, β_{max} =
     0.066 MeV.

     (*i*) Convert 1 mCi to the SI unit of becquerels.

     The answer is 3.7×10^7 Bq (37 MBq). If you got this wrong,
     refer back to 1.4.1.

∏   (*ii*) When purchased, the source had an activity of 1 mCi
           (37 MBq). Calculate the activity 25 years after pur-
           chase.

     The answer is 0.841 mCi (31.12 MBq). If you got this wrong,
     refer back to 1.3.2.

∏   (*iii*) What is an approximate value for β_{mean}?

     The answer is approximately 0.022 Mev. If you got this
     wrong, refer back to 1.2.2.

Open Sources

Sources of this type are mainly used in chemistry and medicine. We
are not directly concerned with the medical applications, although
we shall need to outline the very important area of radioimmunoassay (4.3.4), which is the radioanalytical technique particularly concerned with clinical work. Similarly, many commercially available
labelled compounds can be used in either chemical or clinical research work.

For chemical purposes we can broadly distinguish between the need to use a specific compound labelled with a radioisotope, or more simply to use a solution containing the radioisotope. Put another way, are we following the reactions of a particular compound, or are we simply using the radioisotope as tracer?

Let us take a simple example.

The compound NH_2SO_2OH, (sulphamic acid), hydrolyses slowly to give sulphate ions

$$NH_2SO_2OH + H_2O \rightarrow NH_4^+ + HSO_4^- \rightarrow NH_4^+ + H^+ + SO_4^{2-}$$

This reaction has been used in gravimetric analysis to measure Ba^{2+} by precipitation of $BaSO_4$. The precipitate is obtained in a form that is more easily filtered, and less susceptible to coprecipitation, than is the precipitate obtained by direct addition of H_2SO_4 to solutions containing Ba^{2+} ions.

∏   Can you remember the name given to this technique?

Because the precipitating agent is generated by reaction within the solution rather than by external addition, it is called *homogeneous* precipitation.

Suppose we wish to measure the rate of the hydrolysis reaction ie the rate of formation of SO_4^{2-} ions. We could do this by using sulphamic acid labelled with ^{35}S (ie *a quite specific labelled compound*), and precipitating the $^{35}SO_4^{2-}$ with inactive $BaCl_2$ solution, followed by counting the $Ba^{35}SO_4$. Alternatively we could use inactive sulphamic acid, and precipitate with labelled $BaCl_2$ solution. If we choose the latter it would not much matter which isotope of Ba we use, though the most convenient commercially available one is ^{133}Ba, as a simple solution of $BaCl_2$ in dilute HCl.

It is not practicable to list, or categorise, the labelled compounds and isotopes that are available; indeed the research products and medical products catalogues of Amersham International plc (essential reading if you do become a user of radioactive materials) only differ from 'conventional' chemical catalogues in that the products are radioactive. There is a particularly wide range of compounds containing ^{14}C, reflecting the widespread research interest in organic reaction pathways, in biochemistry, and in molecular biology.

The products are specified in terms of their chemical and radioactive concentration. For compounds, this is most commonly as the source strength expressed in curies (becquerels) per millimole. For simple solutions such as the $BaCl_2$ solution we mentioned earlier it is as curies (becquerels) per mg of the ion concerned. (see 1.4.3)

SAQ 2.1c $^{133}BaCl_2$ solution is available at a specific activity of 370 MBq mg^{-1} Ba. Calculate the disintegration rate of a sample of this solution containing 1 mg of $BaCl_2$, expressing your answer as disintegrations per minute. ($A_r(Cl) = 35.45$; $A_r(Ba) = 137.34$).

Finally you should note that if you do become a user of radioisotopes in the form of open sources you will need to take particular care in relation to the purity, stability, and storage of labelled compounds. We are not simply talking about *chemical* purity; we also need a measure of the concentration of the compound in the correct labelled form (ie with the label in the right position), the concentration of any other labelled compounds present, and so on. Purity takes on an extended meaning!

Also, because the ionising effects of the emitted radiations may become localised in the solids/liquids concerned (see Section 1.2) we can expect significant decomposition, particularly for low energy β^- emitters such as ^{14}C and ^3H. Decomposition rates of 1% per month are not uncommon.

2.1.4. 'Home-based' Preparations

What are the alternatives to buying labelled materials, or to having your own stable samples irradiated? Are there laboratory-based irradiation systems? The answer is that they are very limited. We shall consider them in more detail in Section 4.4; for the present we need only say that the most readily available sources use the nuclear reaction

$$^9_4Be + ^4_2He \rightarrow ^{12}_6C + ^1_0n$$

The α particles (^4_2He nuclei) are provided by isotopes such as $^{241}_{95}Am$ (Am = americium). Unfortunately, the neutron flux produced by such sources is many orders of magnitude lower than reactors, and hence irradiations are limited to materials which particularly readily capture neutrons and for which the product has a very short half-life.

However, there is one way of improving on this situation. When a nucleus captures a low energy neutron it does not absorb all the neutron energy, it emits a γ ray. As it emits the γ ray the nucleus *recoils*, and the recoil energy is often sufficient to break a covalent bond to the atom concerned. The effect is known as a Szilard–Chalmers reaction, after the original discoverers. This can be illustrated by reference to iodoethane:

$$\underset{\substack{\text{captures}\\\text{neutron}}}{\text{H}-\overset{\overset{\text{H}}{|}}{\underset{\underset{\text{H}}{|}}{\text{C}}}-\overset{\overset{\text{H}}{|}}{\underset{\underset{\text{H}}{|}}{\text{C}}}-\text{I}^{127}} \longrightarrow \text{H}-\overset{\overset{\text{H}}{|}}{\underset{\underset{\text{H}}{|}}{\text{C}}}-\overset{\overset{\text{H}}{|}}{\underset{\underset{\text{H}}{|}}{\text{C}}}-\text{I}^{128} \xrightarrow[\text{breaks}]{\text{C}-\text{I}} \text{H}-\overset{\overset{\text{H}}{|}}{\underset{\underset{\text{H}}{|}}{\text{C}}}-\overset{\overset{\text{H}}{|}}{\underset{\underset{\text{H}}{|}}{\text{C}}}{}^{\bullet} + {}^{\bullet}\text{I}^{128}$$

Quite often the bond re-forms, but in this case a significant proportion of the active iodine radicals combine to give labelled molecular *iodine* containing ^{128}I. Note that an *oxidation state change has occurred*, and a competent chemist should be able to devise a separation procedure of reactant and product. Ask yourself how you would determine I_2 volumetrically, and hence in this case perhaps separate the I_2 from C_2H_5I. The answer is with sodium thiosulphate, which gives water-soluble labelled sodium iodide which can be separated from the water-insoluble C_2H_5I (labelled or otherwise) and I_2:

$$^{128}I_2 + 2Na_2S_2O_3 \rightarrow Na_2S_4O_6 + 2Na^{128}I$$

The Szilard–Chalmers reaction is quite widely applicable; for instance coordination compounds have been used as targets.

SUMMARY AND OBJECTIVES

Summary

Synthetic radioactive isotopes are normally made by altering the neutron : proton ratio of stable isotopes, most commonly by bombardment with either neutrons or charged particles. The radioactive isotope is then converted to the appropriate physical and chemical form for eventual usage. A wide variety of radioactive sources and radiochemicals is available, either as sealed or open sources. Very few radioactive isotopes can be conveniently prepared by laboratory-based irradiations.

Objectives

You should now be able to:

- identify the most important factors in preparing radioactive isotopes;

- recognise the importance of reactor irradiation of stable material;

- outline typical processing sequences;

- describe the various types of commercially available product;

- relate the types of source to their potential usage;

- outline a typical laboratory-based irradiation process.

3. Practical Aspects

3.1. RADIATION DETECTION AND COUNTING— GENERAL PRINCIPLES AND GAS IONISATION METHODS

Overview

The section is designed to introduce you to the basic principles of the interaction of radiation with matter, and particularly to emphasise the importance of ionisation. It outlines the most common counting methods, and gives more detailed information on gas ionisation methods.

This topic can be one of great length and complexity; major textbooks have been written on individual techniques of radiation measurement, and these review both the theory and practice of such techniques. For this course it is possible only to give an outline of the important points; if you become involved in experimental radiation work you will certainly require a more detailed programme. The approach adopted here is to indicate briefly what happens when radiation interacts with matter, and then to see how the various effects can be used experimentally to detect the incident radiation.

3.1.1. Principles of Detection

The detection of particles and rays emitted by radioactive isotopes depends upon their energy, their type, and the nature of the material into which they pass.

The important point about the energies of emitted particles or rays involved in radioactive decay is that they are high by comparison with chemical bond energies and ionisation energies of atoms and molecules. Roughly speaking, bond and ionisation energies are some 10^3–10^6 times *less* than radioactive decay energies and thus it is not surprising that *ionisation* of atoms or molecules into which the radiation passes is very important. The extent of the ionisation caused is very dependent on the particle or ray causing it and on the nature of the receiving material, and thus it is again not surprising that the various radiations have different penetrating powers. It is possible to treat the various effects quantitatively, and this is very important in determining appropriate shielding materials (Section 3.5), but for our purposes we can summarise qualitatively.

α particles: very short range—a few millimetres in air, and in-significant in denser materials.

β particles: at low energies comparable to α particles, but other-wise roughly ten times as penetrating as α particles of the same initial energy.

γ rays: except at low energy, γ rays are very penetrating and nor-mally require quite dense solid materials for optimum absorp-tion.

(At this point you should read again the appropriate paragraphs of Section 1.2).

What can we glean from this general information that may be useful in determining a suitable counting method?

(*a*) If ionisation is occurring, then electrons are being liberated, and it should be possible to collect these electrons and relate the magnitude of the effect to the amount of incident radia-tion causing it. Such *ionisation detectors* are fundamental to the counting of activity.

(*b*) The electrons produced when radiation is absorbed by matter will be quite energetic, and we might expect that they will stim-ulate other interesting effects as they lose their energy. We may be able to use such effects in a detector system, and one such method, known as *scintillation counting*, depends on the production of light photons in a suitable absorber.

(*c*) Because of the wide variation in penetrating powers of the various radiations we might predict that the method adopted for counting the radiation from a particular radioisotope normally will be determined by the type and energy of the radiation emitted. A detector that is appropriate for low energy negatrons is unlikely to be useful for high energy gamma rays. To take this a stage further we might expect that detection methods for radiation of very low penetrating power will pose quite major problems. For radiation of medium penetrating power (ie most negatrons) we could expect to use fairly low density materials (usually gases). For radiation of higher penetrating power (most γ rays) we might expect to use much denser materials (usually solids).

(*d*) Given that the effects of radiation are dependent on the energy of the radiation, and that this energy is characteristic of the isotope emitting it, it could well be possible to relate the *magnitude* of the measured effect to the *initial energy* and thus to count only radiation of a given initial energy. This proportionality between initial energy and measured output signal is an important feature of scintillation methods and of some ionisation methods, and enables us to count selectively individual isotopes in a mixture (see later).

We can now consider (in outline only) the most common detector systems, which can be summarised under the heading: gas ionisation systems; solid state (semiconductor) ionisation systems; solid state scintillation counters; liquid scintillation methods.

3.1.2. Gas Ionisation Systems

When radiation passes through a gas, some of the energy associated with the radiation is passed on to some of the gas molecules, which may ionise to a positive ion and an electron. In the absence of any mechanism for collecting the electrons or ions, they will recombine.

However, if a suitable electrical field is applied, the electrons or ions can be collected, and the magnitude of the resulting current can be related to the amount of radiation initially responsible for the ionisation. It is usual for the applied field to be of such a strength that not only are the electrons or ions collected, but that ion multiplication occurs. This means that in being collected, the electrons are accelerated to such an extent that they can cause further ionisation. The major types of gas-ionisation counters are:

(*a*) Gas Flow Proportional Counters

In this type of detector the applied field is such that the number of ion pairs collected is proportional to the energy of the incident radiation. This is achieved by placing the source *inside* the counter, and by carefully controlling such experimental parameters as the geometry of the counter, the flow rate of the gas, and the applied voltage. Because the sample is placed inside the counter problems of penetration of radiation are no longer significant, with valuable consequences for α emitting isotopes (see later).

In principle this an ideal situation; since the output pulse is proportional to the energy of the incident radiation it should be possible, by using a pulse height analyser, to differentiate between pulses of different heights and hence to discriminate between different isotopes. In practice there are experimental difficulties in operating such counters, particularly the need for a very stable high voltage supply and a very stable, high gain, pulse amplifier. The gas used for the counter (eg air-free, dry, 10% methane in argon) is expensive. Thus, although proportional counters are excellent for some specific situations in which a high and reproducible detection efficiency is required (such as calibration of standards) they have not been widely used.

However, you should note that in recent years the need to investigate low levels of radioisotopes in environmental samples, and particularly α emitting isotopes, has led to some increase in usage. This is because the sample to be counted is normally placed *inside* the counter, so that the absorption problems, typical of α emitters, are minimised.

(*b*) **Geiger–Muller Counters** (see Fig. 3.1a)

In this kind of detector the applied field is increased to such an extent that the ionisation induced by the acceleration of the initially produced ions and electrons results in virtually all of the gas in the vicinity of the anode being ionised. (This is known as an 'avalanche' effect). The proportionality of the 'incident energy–output pulse' relationship no longer applies, but the design and ease of operation of Geiger counters is simple and effective.

There are two main types of Geiger counter.

(*i*) End-window counters are normally used for counting solid samples. The counter (Fig. 3.1a) is cylindrical, with a thin mica 'end-window'; the cylindrical case is designed to act as the cathode, and a central metal wire (usually tungsten) is the anode. The counter usually is mounted vertically inside a lead castle (to eliminate external radiation), and the sample to be counted is placed directly under the end-window of the counter. In using an end-window Geiger counter it is necessary to take account of such factors as the distance of the end-window from the source, absorption of the radiation by the window, and scattering of the radiation.

(*ii*) Geiger counters for liquid samples are of glass construction. They consist of a central gas-filled tube containing the electrodes surrounded by a cylindrical 'sheath' into which the liquid to be counted is placed (Fig. 3.1a). Liquid counters have advantages in terms of geometry (the liquid almost entirely surrounds the counter) but absorption of the radiation by the glass may be significant.

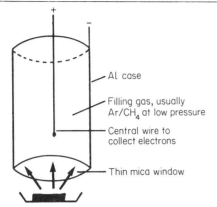

'END WINDOW' GM TUBE

Al case

Filling gas, usually Ar/CH$_4$ at low pressure

Central wire to collect electrons

Thin mica window

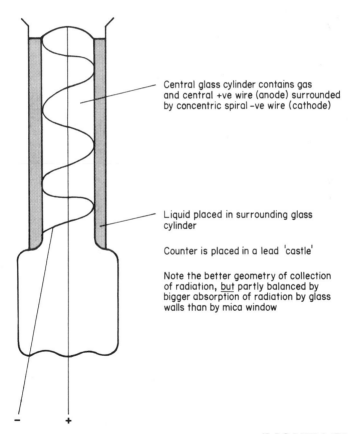

Central glass cylinder contains gas and central +ve wire (anode) surrounded by concentric spiral -ve wire (cathode)

Liquid placed in surrounding glass cylinder

Counter is placed in a lead 'castle'

Note the better geometry of collection of radiation, but partly balanced by bigger absorption of radiation by glass walls than by mica window

'LIQUID' GM TUBE

Fig. 3.1a. *Outline diagrams of Geiger–Muller counters*

Whichever kind of Geiger counter is used it is necessary to establish the operating characteristics (Fig. 3.1b). The optimum voltage for operation depends on the dimensions and geometry of the tube, and the filling gas (normally Ar/CH_4), and is determined experimentally.

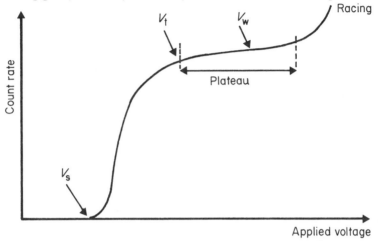

V_s = starting voltage; V_t = threshold voltage;
V_w = working voltage.

Fig. 3.1b.   *Characteristic response curve of a Geiger counter*

You will need practical guidance on the use of a particular Geiger–Muller counter before you are competent to use one.

∏   What are the limitations of Geiger counters? Think of the penetrating properties of radiation and you may predict that there are two types of radiation, particularly, for which Geiger counters are inappropriate, namely

   (*i*) low energy β^- and α particles will not be able to penetrate the mica or glass windows, and hence will not readily be counted, and

   (*ii*) γ rays (particularly of medium to high energy) will pass through the filling gas causing little ionisation, and hence will give a low efficiency.

Thus a different means of counting is necessary in these situations.

SAQ 3.1a

The following isotopes emit the radiation(s) specified:

(*i*) ^{14}C, low energy β^-

(*ii*) ^{60}Co, high energy γ and medium energy β^-

(*iii*) ^{239}Pu, medium energy α

(*iv*) ^{32}P, high energy β^-

Which two of these isotopes can most usefully be measured by proportional counters?

(*i*) and (*ii*)

(*ii*) and (*iii*)

(*i*) and (*iii*)

(*iii*) and (*iv*)

(*ii*) and (*iv*)

Which of the isotopes is best counted by end-window Geiger counting?

SUMMARY AND OBJECTIVES

Summary

The choice of counting methods depends primarily on the type and energy of the emitted radiation. Because radiation causes ionisation to occur in materials through which it passes, many radiation detectors use the measurement of the amount of ionisation as a means of measuring the incident radiation. Such detectors include proportional counters and the widely used Geiger counters.

Objectives

You should now be able to:

● appreciate the complexities of the interactions of ionising radiations with matter;

● recognise that ionisation of molecules and atoms in material absorbing the radiation is very important;

● relate the type of radiation, and its energy, to likely counting techniques;

● explain the basic principles of proportional and Geiger–Muller counters, and identify the limitations of these two counting methods.

3.2. RADIATION DETECTION AND COUNTING— SCINTILLATION METHODS

Overview

The section is designed to introduce you to the common features and functional parts (particularly photomultiplier tubes) of scintillation counters. It surveys the various types of scintillator, and relates these to particular areas of application. The section also describes some important experimental features of the methods.

In the previous section you saw that the interaction of radiation with matter causes ionisation, through the deposition of energy in the absorbing medium. Atoms or molecules in the absorber will thus become excited, either by direct interaction with the radiation or by interaction with electrons produced in a previous interaction. Scintillation counting makes use of the fact that when some materials de-excite they emit a light photon in a wavelength range that is conveniently detected, and thus it should be possible to design a means of measurement based on this effect.

3.2.1. Photomultiplier Tubes

It is insufficient simply to *detect* the light photons emitted by the absorbing atom or molecule; before this can be counted it must be converted to a convenient pulse of electrons. This is achieved in a photomultiplier tube (PM tube) which is an essential component of all scintillation counters.

The first requirement of a PM tube is the ability to convert the light photon produced by the scintillating material to an electron in the tube. This is achieved by having a light sensitive cathode, which consists of a thin film of an alkali metal (usually caesium) deposited on the inside surface of the glass (or silica) face of the PM tube, often called the 'window'. The light photons stimulate the caesium to emit electrons which are accelerated through a potential of some 200 V to strike another electrode. This electrode, made of a beryllium alloy, is called a 'dynode'. When an accelerated photoelectron strikes the dynode the latter emits an average of four electrons. These in turn are accelerated towards the next dynode where the same effect occurs. Typically there are eleven dynodes, so that the electron multiplication is of the order of 4^{11}, or roughly 4×10^6, depending on the operating conditions. These electrons are collected at an anode, and the resultant pulse is fed to an amplifier circuit which typically consists of a preamplifier and a linear amplifier.

It is possible to show that using appropriate circuitry the height of the output pulse is directly proportional to the energy of the incident radiation, and this is fundamentally important in γ ray spectrometry (see Section 3.4).

3.2.2. Scintillation Counting with Solid Scintillators

You have seen previously that the most significant penetration into solid materials is achieved by γ rays; furthermore, because they cause so little ionisation in gases they are only poorly measured by gas ionisation methods. Hence it is perhaps not surprising that γ rays are most efficiently counted by using a solid absorber that can act as a scintillator. This is normally called a *crystal* or a phosphor; by far the most popular crystal is sodium iodide containing traces of thallium(I) iodide, usually written as NaI(Tl). The detailed mechanism by which the crystal emits a light photon is not important here; broadly speaking, the TlI introduces dislocations into the NaI lattice, and it is from these dislocations (or 'activator centres') that the photons are emitted.

Crystals are available in many sizes; typically they are cylindrical and encased in a light-tight aluminium can coated on the inside with a reflective layer of magnesium oxide. One end of the crystal is left clear (the 'window') and this makes contact with the end of the PM tube at which the photocathode is located; contact is facilitated and reflection losses minimised by a film of transparent silicone oil.

A typical arrangement is shown in Fig. 3.2a.

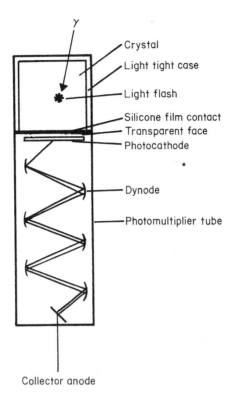

Fig. 3.2a. *Simplified schematic diagram of crystal and PM tube (Electrical Circuitry and all dynodes are not shown)*

Finally you should be aware that although NaI(Tl) is widely used other solid scintillators are available. The main criteria for a scintillator are:

(*a*) good absorption of incident radiation, which in practice means reasonably high density;

(*b*) high efficiency of photon production;

(*c*) little or no reabsorption of the photons by the crystal;

(*d*) emitted photons must have a wavelength compatible with the PM tube (approximately 320–750 nm with a glass window).

In practice some organic compounds meet these criteria, notably anthracene and stilbene. Plastic scintillators are also available. Some relevant data are given in Fig. 3.2b.

	NaI(Tl) >	Anthracene >	Stilbene >	Plastics
Density/g cm^{-3}	3.67	1.25	1.16	1.05
λ_{max}/nm	410	440	410	350–450

Fig. 3.2b. *Some properties of selected scintillators (in order of decreasing pulse height for given radiation)*

Some of these other scintillators are used for detection of radiation other than γ rays; thus, plastic scintillators are frequently used for β^- counting. Also there are some specialised scintillation counters (usually based on NaI(Tl)) for low energy γ and X-ray counting.

3.2.3. Liquid Scintillation Counting

Gas ionisation counters and solid scintillation counters enable us to count isotopes which emit gamma rays and/or medium to high energy β particles. However there are several low energy β^- emitters for which these techniques are inappropriate, and unfortunately these include some of the most widely used isotopes including:

	^{3}H	^{14}C	^{35}S	^{125}I
E_β/MeV	0.015	0.156	0.167	0.035(E_γ)

It is particularly important to have available a counting method capable of routine use for these isotopes because of their value in radioimmunoassay (Section 4.3).

Because of the low penetration of the β^- particles it is essential that the isotope to be counted should be intimately mixed with the scintillator and not external to it as for solid scintillators. This is achieved by dissolving the sample wholly or partly in a solvent containing the scintillator. This technique of liquid scintillation counting is now highly sophisticated, and has been the subject of several textbooks.

The most important components are as follows.

(*i*) Solvent

This is most commonly an aromatic hydrocarbon, typically toluene or xylene; in some cases dioxan is used. Apart from functioning as a solvent, the primary effect is that incident radiation can readily form electronically excited solvent molecules. These readily transfer their energy to the scintillator.

(*ii*) Scintillator

This is often called the *primary solute*. It readily accepts energy from the solvent, is raised to an excited state, and then de-excites by emitting a light photon in a suitable wavelength range (300–400 nm) to be 'seen' and counted by a PM tube. Typically these are fairly large organic molecules.

p - terphenyl (Prr)

1, 4 - diphenyloxazole (PPO)

1, 4 - diphenyloxadiazole (PPD)

(*iii*) Secondary Solute

In some cases where maximum efficiency is required, a secondary solute (scintillator) is added. The excitation energy of the primary solute molecules is transferred to the larger secondary solute molecules which then emit the energy more efficiently and at a longer wavelength, typically 400–500 nm, where the photomultiplier is more sensitive.

A typical secondary solute is

1, 4 – di – 2 – (5 – phenyloxazole) (POPOP)

Although a detailed treatment of liquid scintillation counting is outside the scope of this course it is appropriate to outline some practical points.

(*a*) The optimum situation is where the sample readily dissolves in the solvent (eg some organic compounds in toluene) or is miscible with it (eg aqueous samples with dioxan). However, it is sometimes necessary to add a solubliser that as far as possible does not interfere with the scintillation process. Ethanol is often used, and there are specialised procedures for particular sample types.

(*b*) Where complete solubility or miscibility is unobtainable it is possible to count a gel or emulsion, the emulsion being stabilized by a suitable detergent. Such systems can accept up to 40% aqueous material, which may be very important in the analysis of clinical samples.

(*c*) In practice it is usual to buy a ready-made scintillation 'cocktail' appropriate to the material being counted.

(*d*) It would be expected that for such low energy emitters the pulse height obtained will be quite low, and it may be difficult to separate the signal pulses from thermionic 'noise'pulses usually from the photocathode of the PM tube. Thus, it is common practice to use 'coincidence' counting; two PM tubes are coupled through a coincidence unit

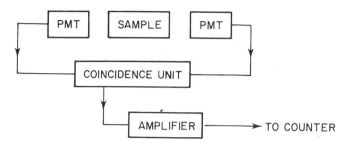

TAKE IN UNUMBERED FIGURE

The coincidence unit passes to the counter only those pulses which arrive virtually coincidentally ($< 1 \mu s$) from each PM tube. Any pulses arising singly are rejected as 'noise'.

(*e*) Liquid scintillation counting is very amenable to the counting of a large number of samples, and modern instrumentation frequently incorporates automatic sample changers and microprocessor control.

(*f*) Frequent recalibration and measurement of counting efficiency is necessary.

SAQ 3.2a

The following is a list of key words and compounds from this section. Place them in the correct pairings.

thallium(I) iodide; secondary solute; dynode; magnesium oxide; anthracene; caesium film; lattice dislocation; beryllium alloy; primary solute; photocathode; 1,4-di-2-(5-phenyloxazole); p-terphenyl; reflective layer; organic crystal.

SUMMARY AND OBJECTIVES

After they have been excited by incident radiation some materials de-excite by a mechanism involving emission of a photon of light. This flash of light can be multiplied to give a pulse of electrons which can be counted. The scintillation material can be either a solid (most commonly used for γ ray counting) or a liquid (most commonly used for counting of low energy negatrons).

Objectives

You should now be able to:

- explain the basic principles of scintillation counting, and differentiate between solid and liquid scintillation counting;

- describe the important components of scintillation counters, and identify the main experimental features of scintillation counting.

3.3. RADIATION DETECTION AND COUNTING — SEMI-CONDUCTOR Ge(Li) DETECTORS

Overview

In this section the two most important types of semiconductor are briefly described, with an indication of how they can be combined in a radiation sensitive device. The section considers the limitations and advantages of such detectors in relation to other more traditional detectors.

In the preceding sections you have seen how solid materials can act as scintillation detectors, and how gases can act as the basis of ionisation counters. In recent years, as a result of advances in semiconductor technology, several new solid state detectors have been developed which effectively function as ionisation counters. In this section we shall consider the background to counters of this type which are particularly useful for γ ray counting: the so-called lithium drifted germanium detector, which is usually abbreviated to Ge(Li) — pronounced 'jelly'!

3.3.1. Theoretical Basis of Ge(Li) Detectors

It is necessary to briefly review simple semiconductor theory with respect to germanium, a group IV element. If a small amount of a group V element (eg phosphorus) is introduced into germanium there will be occasional sites where four of the electrons (of the P) are bonded, but the fifth is free to migrate. This is known as an *n-type semiconductor*; it contains negative sites (electrons) available for conduction.

Conversely, if a group III element is introduced into germanium there will be sites where only three electrons are available for bonding, ie there will be a deficiency of electrons, or what is known as a 'positive hole'. This is called a *p-type semiconductor.*

The other requirement for a solid state ionisation detector is an element whose atoms are not only small enough to fit into the interstitial sites of Ge, but also are readily ionised. Such an element is lithium.

To construct a Ge(Li) detector the Li atoms are allowed to migrate or 'drift' to some of the p-sites in a cylinder of p–n germanium. Remember that each p-site is deficient in one electron, and each Li is readily ionised and gives up an electron. Thus we reach a situation (Fig. 3.3a) where we have a p-type semiconductor and an n-type semiconductor (effectively the electrodes) separated by an ultrapure and electrically neutral region (called the 'intrinsic region'). If radiation enters the neutral region it will cause ionisation (effectively it is operating like the gas filling of a gas ionisation detector), and by applying a suitable potential the electrons can be collected and amplified to a measurable pulse. Furthermore, the pulse height obtained is directly proportional to the energy of the incident (gamma) radiation, so that we again have the opportunity to discriminate between isotopes emitting radiation of different initial energy.

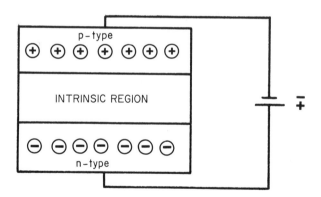

Fig. 3.3a.   *Basic arrangement of a Ge(Li) counter*

3.3.2.   Experimental Features of Ge(Li) Detectors

There are several practical features of Ge(Li) detectors to take into account in considering their use for a given application.

(*a*) As we have noted, lithium atoms are very mobile, and once the Ge(Li) detector has been produced it has to be maintained at a low temperature to prevent further migration of lithium atoms and to minimise thermal excitation of the intrinsic region. In practice this means cooling the detector (continuously—not just when in use) to liquid nitrogen temperatures.

(*b*) Partly because of the small size of semiconductor detectors they have much poorer detection efficiencies than NaI(Tl) crystals. (There are other reasons, beyond the scope of this course).

(*c*) Ge(Li) detectors are not as effective as NaI(Tl) crystals for samples with a high count rate.

(*d*) Ge(Li) detectors must be contained in a vacuum, whereas there are no such restrictions on NaI(Tl) crystals.

These are all quite serious disadvantages. Why then are Ge(Li) detectors so useful? The answer is that they have *excellent resolution*, sometimes approaching 100 times better than NaI(Tl) detectors. What does this mean in practice? We shall briefly consider γ ray spectrometry in 3.4; for the present we only need to know that the physical result as presented on the recorder of the detection system is that each γ ray gives rise to a photopeak, the γ spectrum of a mixture of isotopes being quite similar visually to, for example, the gas chromatogram of a mixture of compounds, or the XRF spectrum of a mixture of elements. The resolution is defined as the full width of the photopeak at half peak (ie maximum) height (FWHM), divided by the peak position (Fig. 3.3b).

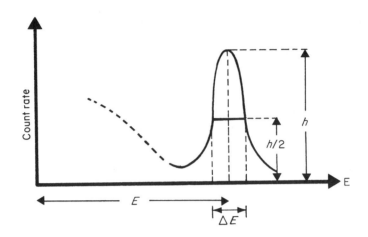

Fig. 3.3b.   *Resolution of a photopeak*

The resolution defined as: $\Delta E/E$, is normally expressed as a percentage. Effectively, resolution is a measure of the minimum energy difference that must exist between two gamma rays for their corresponding photopeaks to be resolved. For comparison between detectors their resolution is usually quoted with reference to a standard γ ray, most commonly the 0.662 MeV γ ray of ^{137}Cs.

For a sodium iodide detector a typical resolution is 8%. Hence at $E = 0.662$ MeV (662 keV)

$$\Delta E/662 \quad = \quad 0.08$$

and $\Delta E \qquad = \quad 53$ keV (0.053 MeV)

For a Ge(Li) detector a typical resolution is 0.3%. Hence at $E = 0.662$ MeV the value of $\Delta E = 2$ keV (0.002 MeV).

The consequence of this dramatic difference in resolution is that Ge(Li) detectors can resolve much more complex mixtures of γ photopeaks than can NaI(Tl) detectors. Typical spectra from the two methods of detection are shown in Fig. 3.3c.

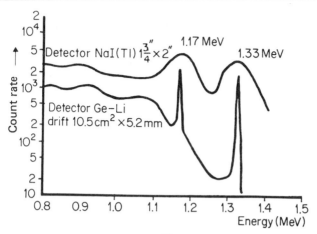

Fig. 3.3c. *Energy spectrum of ^{60}Co obtained with scintillation and Ge(Li) detectors. (After J A Oosting, in* Nuclear Chemistry, *Choppin & Rydberg, Pergamon Press, 1980.)*

SAQ 3.3a The table below gives E_γ values (keV) for four isotopes. Calculate ΔE values, and complete the table. You should use resolutions of 8% and 0.3% for NaI(Tl) and Ge(Li) respectively. From your calculated ΔE values, decide which *two* pairs of gamma rays could *not* be resolved satisfactorily by a NaI(Tl) detector. Can they be resolved satisfactorily by a Ge(Li) detector?

ΔE/keV

Isotope	E_γ/keV	Ge(Li)	NaI(Tl)
^{198}Au	412		
^{69}Zn	440		
^{76}As	559		
^{122}Sb	564		$\longrightarrow$

SAQ 3.3a

In conclusion we can say that if it is necessary to count a single γ emitting isotope, or a simple mixture of γ emitters whose photopeaks are well separated, then the efficiency of a NaI(Tl) detector will be preferred. However, if a complex mixture of γ emitters is present (as in some activation analysis procedures) and if a chemical separation is to be avoided, then a Ge(Li) detection system will be necessary.

SUMMARY AND OBJECTIVES

Summary

Semiconductor detectors are exemplified by the Ge(Li) system. Detectors of this type consist of an n-type semiconductor and a p-type semiconductor separated by an ultra pure, radiation-sensitive region. Despite demanding operating conditions (low temperature, high vacuum) and relatively low efficiency, such detectors are very useful because of their extremely good resolution.

Objectives

You should now be able to:

- distinguish qualitatively between an n-type and a p-type semiconductor;

- explain why a device prepared from the two types of semiconductors can be useful in radiation counting;

- relate the properties of such detectors to other detectors used for the same purpose.

3.4.   RADIATION DETECTION AND COUNTING—γ RAY SPECTROMETRY

Overview

This section describes, in outline, the theoretical background to the interaction of γ rays with matter, and indicates how this frequently leads to very complicated spectra. It also describes the experimental methods for obtaining and interpreting the spectra.

In earlier sections you have seen that for the counting of isotopes which emit γ rays it is possible to use either NaI(Tl) scintillation detectors or Ge(Li) semiconductor detectors, depending on the complexity of the sample. You have also seen that for both types of detector the energy of the initial γ ray is proportional to the height of the electrical pulse that it produces in the detection system. So far we have assumed that:

(a) the γ rays emitted by such isotopes lead only to photopeaks characteristic of the initial energy, and

(b) that experimentally we can normally distinguish such photopeaks.

It is now necessary to investigate the validity of the first assumption, and to accept that a typical γ ray spectrum is likely to be far more complicated than the decay scheme might suggest. Secondly it is necessary to indicate how γ ray spectra can be obtained experimentally.

3.4.1.   The Complexities of γ Ray Spectra

When γ rays interact with a detector, whether it is a scintillation crystal or a semiconductor device, they are interacting with a solid material. What are the factors that can affect these interactions? Clearly there are two primary variables, namely: the energy of the γ ray, and the nature of the solid material. We are particularly concerned with the former. There are three main ways in which γ photons lose their energy on interacting with matter:

(*a*) at low energy the entire γ photon energy is transferred to an electron, which is ejected as a photoelectron;

(*b*) at higher energies part of the energy is transferred to the ejected electron, and the rest is scattered as a lower energy photon (the Compton effect);

(*c*) at still higher energies (>1.02 MeV) the production of a 'positron–electron' pair takes place, and this pair subsequently annihilates to give two γ rays each of energy 0.51 MeV.

Other complicating features of γ ray spectra are

(*d*) if an isotope emits more than one γ ray the energies of the γ photons may in some cases be additive, so that 'sum peaks' are observed;

(*e*) γ emitting isotopes normally also emit β^- or β^+ particles, and when these charged particles lose their energy electromagnetic radiation is produced, which will also show up in the γ spectrum;

(*f*) scattering of the γ photons by materials other than the crystal, such as shielding materials.

If you are sufficiently interested in these aspects and wish for further information you are recommended to read the appropriate parts of the books in the Study Guide (Faires and Boswell; Malcolme-Lawes).

Notwithstanding these complexities it is normally possible in a given spectrum to identify the photopeaks corresponding to the original γ photons, and by using standards to set up a calibration between the pulse height and the γ energy, of the type shown in Fig. 3.4a.

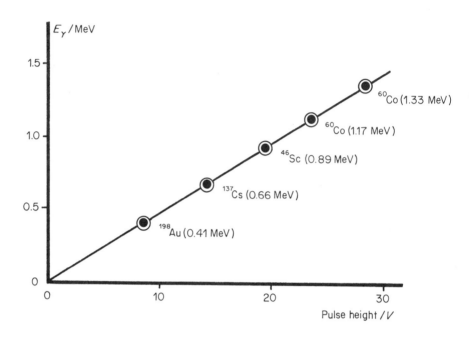

Fig. 3.4a.   *Energy calibration for γ rays*

Using such a calibration it is then possible to identify unknown components of a mixture, eg after neutron irradiation for activation analysis purposes.

3.4.2. Pulse Height Analysis

Fig. 3.4a presupposes:

(*a*) that there is a direct relationship between pulse height and E_γ, and

(*b*) that pulses of different heights can be distinguished.

The proof of the first of these points is not included here, but if you wish to pursue this aspect there are good accounts in the more advanced textbooks.

In its simplest form pulse height analysis consists of arranging electronically to count only those pulses which are above a predetermined pulse height (the lower discriminator) but below another predetermined pulse height (the higher discriminator). The voltage gap between the discriminator levels is known as the 'channel width'. Modern instruments, called multichannel (pulse height) analysers (MC[PH]A), are designed so that the normal E_γ range, often taken as 0–2.0 MeV (0–2000 keV) is subdivided in voltage terms into a certain number of channels—usually multiples of 100, eg 200 or 400 channel analysers, or of 128, eg 256, 512, or 1024 channel analysers. Counts of a given pulse height are stored in the appropriate channel; clearly in principle the more channels, the better, although it should be recognised that increase in number of channels is reflected in an increase in the purchase price of the analyser.

The most important practical point is that the superb resolution of Ge(Li) detectors virtually demands a high quality MCPHA; the poorer resolution of NaI(Tl) crystals is less demanding.

You may also expect that because of the sophistication of an MCPHA system, and the amount of data to be collected and stored, microprocessors are widely used in γ spectrometry (Fig. 3.4b).

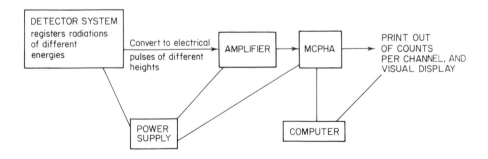

Fig. 3.4b. *Simple block diagram of a γ spectrometer*

See also Subsection 4.4.4 for more on computing inputs to activation analysis.

SUMMARY AND OBJECTIVES

Summary

γ ray spectra are normally far more complicated than might be inferred from a simple tabulation of the decay scheme, principally because of secondary effects when the γ rays interact with the detector. However, because of the proportionality between the energy of the incident, E_γ, and the pulse of the resulting photopeak, and by using modern pulse height analysers it is normally possible to distinguish and count individual γ emitting isotopes in a mixture.

Objectives

You should now be able to:

- appreciate the complexities of γ ray interactions with solids, and relate these to the corresponding γ ray spectra;

- describe the experimental basis for discriminating γ rays of different initial energy.

3.5. STATISTICS OF COUNTING OF RADIOACTIVE SAMPLES

Overview

This section is designed as a simple introduction to the statistics of counting of radioactivity, and to place the relevant factors in the wider framework of statistics in analytical chemistry. It outlines some of the consequences of a statistical treatment in relation to experimental procedures.

Quantitative chemical analysis essentially depends on the ability to measure the magnitude of a chosen property of the system being studied, and then to relate the measured value (however it has been obtained) to the quantity of material giving rise to the effect or property being measured. It is very tempting to give the value obtained ('the result') some *absolute* credibility—calculate an *exact* value of concentration from for example a titre, an optical absorption reading, or a measurement of emf. Your practical experience, whichever techniques you are most familiar with, should at least have taught you to be cautious in converting experimental data to absolute values of concentration. There are numerous sources of error in any analytical determination; the magnitude of these errors will vary both individually and cumulatively, and it is essential for an analyst to make a sensible assessment of the validity of the results obtained. The use of highly sophisticated instrumental methods with more rapid accumulation of data, ease of repetitive measurements, and storage and computer treatment of data, has placed even greater emphasis on the correct use and interpretation of analytical results.

The purpose of this section is to discuss briefly particular aspects of the measurement of radioactive samples. The more widely applicable aspects of statistics and data treatment in analytical chemistry are considered elsewhere in the ACOL programme, and in all standard textbooks on analytical chemistry. It is assumed that you are familiar with such basic concepts as mean, standard deviation, relative standard deviation, variance, and distribution functions. In the text which follows the important practical aspects of statistics in radioactive measurements are given in outline only; their derivations from the wider theoretical concepts are not considered.

You can refresh your knowledge of basic concepts by attempting
SAQ 3.5a.

SAQ 3.5a   In an experiment to check that a Geiger counter
was working the following 20 sets of counts were
obtained, in each case over a 30 second period.

9843, 9774, 9858, 9828, 9831, 9768, 9927, 9834,
9792, 9804, 9819, 9882, 9846, 9816, 9696, 9897,
9687, 9876, 9789, 9858.

(i)   Calculate the mean value of these results,
$\bar{x}$.

(ii)   Obtain an estimate of the true standard deviation, s.

(iii) Calculate the relative standard deviation
(coefficient of variation).

(iv)   Check how many individual results lie in
the range $(\bar{x} - s)$ to $(\bar{x} + s)$.

3.5.1. Decay Rate and Statistics

It is appropriate at the beginning to remind yourself that radioactive decay is a random process. You are strongly advised to look back at Section 1.3, and refresh your memory on such aspects as the fundamental decay law, decay constant, and activity.

Because the activity of a sample is proportional to the number of radioactive atoms present, you might perhaps think that statistics will play a relatively simple part in radioactive counting. However, before we even consider the practical problems it is not difficult to use basic theory to put the problems in perspective.

Let us suppose that in a simple tracer experiment we are required to count a sample containing 3.7×10^3 Bq of radiation. In a counter of (say) 10% efficiency this would give a measured count rate of $0.1 \times 3.7 \times 10^3 \times 60$ counts min^{-1}, ie 2.22×10^4 counts min^{-1}. This sounds a lot, but what proportion of the radioactive nuclei are actually decaying during the course of a counting period of one minute?

You should recall the fundamental decay law:

$$dN/dt = -\lambda N \qquad\qquad (1.3a)$$

where $\qquad \lambda = 0.693/t_{0.5}$

Hence the number of radioactive nuclei N is given by:

$$N = -(dN/dt) \times (t_{0.5}/0.693)$$

Let us assume that our sample contained ^{32}P, for which $t_{0.5} = 14$ days. Since we are expressing $-dN/dt$ in becquerels (ie disintegrations s^{-1}) $t_{0.5}$ must be expressed in seconds also

$$\therefore \quad N = 3.7 \times 10^3 \times 14 \times 24 \times 60 \times 60/0.693$$

$$= 6.458 \times 10^9 \text{ nuclei of } ^{32}P$$

Hence the percentage of nuclei decaying in a counting period of one minute is $3.7 \times 10^3 \times 60 \times 100/6.458 \times 10^9$

$$= 3.44 \times 10^{-3}\%$$

This of course is a very small proportion of the radioactive nuclei present; since such a small change is occurring in the property which is the basis of our measurements, it is not surprising that it is important to use statistics properly in the measurement of radioactivity.

It is completely wrong to think of activity, and of measured count rates, as having some *absolute* value. You should be clear that in measuring the activity of a radioactive sample there will always be a *natural variation about a mean value*. We can amplify this point much more satisfactorily in the practical work associated with this Unit.

3.5.2. Practical Considerations

If you are involved in radioactive work on a regular basis it is necessary to test your results frequently for their statistical significance, using for example least squares analysis and control charts, and checking for non-random errors. The latter can be particularly important since, as you have already seen, it is in the nature of measuring radioactivity there is always a random error in the result. However, for this Unit we shall only survey some simple practical points.

(a) Optimum Counting Rate

If you are carrying out a volumetric analysis you can normally control the concentration of the titrant so that the volume used is neither so large that you need to refill the burette nor so small as to be subject to a proportionately large error. Similarly in radioactive work the sample should as far as possible not have a very low count rate, nor should it be excessively high. In the latter situation, not only could the count rate be outside the working range of the counter, but a hazard to the operator could be introduced.

However, there is another constraint that derives from the statistics of counting radioactivity. Designers of radioactivity experiments for students often try to arrange for the samples being measured to have an experimental count rate of about 10^4 counts min^{-1}. In a less controllable situation (for instance an analytical process using different sample sizes) it is unlikely to be possible to arrange this, but nevertheless it is common practice to obtain at least 10^4 counts from the source, however long it takes.

What is the significance of 10^4 in this context?

If the total observed count is 10^4, the standard deviation (ie $\sqrt{\text{count}}$) is 10^2 and the random error with a 68.3% confidence limit is 1%. If you have a good knowledge of statistics you will appreciate that this derives from the mathematics of normal distribution. If you do not have such a knowledge you will have to accept that *a total count of 10^4 reduces the random error to 1%* (which is reasonable for many experiments), and that even with 10^4 counts there is roughly a one in three chance that this limit will be exceeded.

(*b*) Background Counts

Because of natural radioactivity (eg from cosmic radiation) there is always a 'background' count associated with measuring a radioactive sample. This is normally reduced to a minimum, for instance by placing sample and counter inside a lead castle. In this situation the background should be low, say 30 counts min^{-1}. To accumulate a total of 10^4 counts would take roughly 5–6 h; clearly if one is measuring a sample giving 10^4 counts min^{-1} it would be stupid to spend 5–6 h making sure that the statistics of counting the background are satisfactory! The moral should be obvious; the background count will become more significant the nearer it is to that of the sample. It is possible to show that the optimum counting times for the sample (t_s) and the background (t_b) are related to the count rate of the sample and the background by the expression

$$t_s/t_b = \sqrt{\frac{\text{count rate of sample}}{\text{count rate of background}}}$$

SAQ 3.5b   Preliminary experiments show that the count rate of a sample is 4860 counts min^{-1} and that of the background is 60 counts min^{-1}. You have 20 minutes available for counting both sample and background for periods which optimise the statistics for each. What are the best values of t_s and t_b?

(*c*) Dead Time ('Paralysis Time')

In the experimental work associated with this Unit you will use a Geiger counter. Because of the way such a counter functions it does not register all the counts that should theoretically correspond to the ionisations of the filling gas of the counter. There is what is known as a 'dead time' or 'paralysis time', during which the counter (having just registered a count) is discharged and will miss an ionisation occurring whilst it is discharged. The counts which are lost increase with the count rate, and must of course be taken into account. Experimentally the dead time is controlled electronically to a predetermined value (often 100 μs) and corrections must be made using data tables relevant to the chosen dead time.

(*d*) Instrumental Errors

Inevitably counting apparatus will occasionally malfunction. An amplifier in an electronic circuit may be affected by, for instance, sources of microwave radiation; an automatic timer may not work properly. Such faults will give rise to non-random errors, which because they may be intermittent, may not be easily recognised. There are special tests for non-random errors, and these should be carried out regularly.

SUMMARY AND OBJECTIVES

Summary

The application of simple statistical considerations to treatment of measurements of radioactivity is very important, and leads to the need to assess optimum counting rate, the effect of background counts, and effects deriving from the counting apparatus itself.

Objectives

You should now be able to:

- recognise that radioactivity is a random process;

- appreciate that it is necessary to treat the counting of radioactive samples statistically;

- explain the simple statistics of decay rate;

- identify the important practical aspects of statistics of counting.

3.6. RADIATION PROTECTION AND CONTROL

Overview

This section is designed to introduce you to the basic principles of radiation protection. It describes the different ways in which dose rates can be calculated, and outlines the most common ways of achieving protection. Finally the section discusses radiation monitoring systems.

Because of the potential hazards of radioactive materials their use in all working situations is very closely controlled. In the UK this is generally exercised through the Radioactive Substances Act of 1960, and numerous subsequent regulations designed to cover specific situations. At the international level the whole topic of radiation protection is covered by the International Commission on Radiological Protection (ICRP). Their recommendations, though not mandatory on individual countries, nevertheless are generally accepted and in the UK form the basis of current working practice. The UK organisation with particular responsibilites is the National Radiological Protection Board (NRPB). Many aspects of radiation hazard control, and the associated administrative procedures, are embodied in the Health and Safety at Work Act of 1974, which is enforced through the Health and Safety Commission.

You should be absolutely clear that if you find it necessary to work with radioactive materials both you, and your working situation, will be subject to very stringent monitoring and control. It will be essential for you to receive training in all relevant aspects of radiological safety and good working practice. The whole area of radiation protection and 'health physics' is highly specialised, and you will require expert advice. The material which follows is NOT designed with this in mind; it is simply an indication of the kind of factors which are the bases of regulatory procedures.

3.6.1. General Background to Radiation Protection

There are two ways in which individuals are subject to radiation, by exposure to external sources, and by radioactive materials becoming absorbed into the body. Before we consider the implications of these points you should be clear that everyone, whether working with radioactivity or not, is exposed to radiation. Data are published periodically; the most recent figures published by the NRPB show that of the total annual dose from all sources to the average member of the population, 87% arises from natural background (cosmic rays, rocks and soil, the atmosphere [radon and thoron], and the human body). Of the man-made part, the largest contribution (11.5%) is from medical exposures, and particularly X rays. The kind of work we are concerned with in this course makes a negligible contribution.

We are not concerned with hazard control as it may affect the population at large; there has been a general tightening of pollution control, and you might guess that organisations working with radioactive materials are subject to very stringent controls on the amount of activity they can put into the natural environment. There are many highly sophisticated analytical procedures for monitoring such effluents.

Our concern is essentially with laboratory hazards. Exposure of workers to external sources can arise either *without* contact with the source (eg if a source is being manipulated with handling tongs), or *with* contact (eg if contact is made with a contaminated working surface or apparatus). Thus we shall need some information on absorption of radiation by materials, the use of shielding materials, the effects of distance, the relationship between exposure dose and absorbed dose, monitoring of surfaces, and of personnel, and many other factors.

The absorption of radiation into the body will principally be orally or by breathing contaminated air, although there is a possible route through contact of activity with cuts in the skin. For these circumstances the main requirements will be the design and implementation of good working practices, carefully thought out experimental procedures, and monitoring of laboratory (airborne) radiation.

3.6.2. Dose Rates and Relevant Limits

When radiation travels through any medium it causes ionisation, the extent of which depends on the nature of the particle or ray concerned (Section 1.2), and it transfers energy to the medium. The unit of *exposure* is expressed in terms of the ionisation produced, or the *quantity of electrical charge produced*; the latter is particularly useful, since most monitoring instruments function by measuring the resultant electrical current. The unit is the ROENTGEN, defined as

$$1 \text{ roentgen} = 2.58 \times 10^{-4} \text{ C kg}^{-1}$$

Strictly this is the pre-SI unit; on the SI system the unit (not at present named) is the amount of radiation that produces 1 C kg^{-1}. For normal work this is impractically large, and seems unlikely to be widely used. Strictly speaking, both units apply only to γ and X radiation.

Units of *absorbed* dose are defined in terms of the *amount of energy deposited*

$$\text{Pre SI} : \text{the rad, where 1 rad} \quad = \quad 10^{-2} \text{ J kg}^{-1}$$

$$\text{SI} : \text{gray(Gy), where 1 gray} \quad = \quad 1 \text{ J kg}^{-1}$$

Hence 1 gray (Gy) = 100 rad

The relationship between exposure and absorbed dose is complex, but because experimentally it is much easier (and more common) to measure exposure dose it is necessary to have a 'rule of thumb' equivalence. This is that for soft tissue

$$1 \text{ roentgen} = 0.87 \text{ rad}$$

It is important to note that if an exposure dose is measured experimentally it can be equated to the same absorbed dose with a significant safety margin.

Next we need to consider the concept of '*dose equivalent*'. The biological effects of radiation depend not just on the *total* energy deposited, but on the *extent* of its deposition as a function of the *distance travelled*, which depends on the nature of the particle or ray. Thus there is introduced the concept of the 'quality factor' Q; this attempts to quantify the biological effects of different types of radiation. The higher the quality factor the greater is the biological effect. Like the writer, you may find this a rather perverse use of the word 'quality'! In practice, the dose equivalent is calculated as the absorbed dose multiplied by the quality factor:

Pre SI : Dose Equivalent in REM   =   Abs. Dose (Rad) × Q

  SI : Dose Equivalent in sievert   =   Abs. Dose (Gy) × Q

Since Q is dimensionless, the units of dose equivalent are the same as those of absorbed dose:

1 sievert (Sv) = 1 J kg^{-1}

Some typical Q values are:

X rays; γ rays; $Q = 1$

fast neutrons; protons $Q = 10$

α; fission fragments $Q = 20$

Finally we should note that some attempt is made to quantify the non-uniform internal dose effect, by introducing a 'distribution factor', but since at present this is taken as one, it does not affect calculations.

Present regulations relate dose equivalent, through a series of complicated assumptions/relationships on risks, to limits on exposure. In its simplest form, for 'whole body' irradiation of workers with radiation, the *annual dose-equivalent limit* should be 50 mSv (millisievert), and for the population at large 5 mSv.

For setting limits on *intake* of radioactive materials the situation becomes even more complicated, since various organs have different retention times, certain elements are retained fairly specifically by some organs, some compounds of an individual element are more soluble in the body than others etc. Thus the ICRP makes recommendations on such factors as the 'Annual Limit of Intake' (ALI) for the inhalation or ingestion of individual radioisotopes in each of several chemical forms. By making assumptions about such things as body weight, breathing rate, and working hours it is then possible to calculate the 'Derived Air Concentration' (DAC). Tables of values of these parameters are available for radiation control.

You should also have noted that in the definitions and discussion so far there has been no mention of *time*. It will of course be necessary to take account not just of the total dose received, but of the time over which it was received, and this will be particularly important in designing experimental procedures where exposure to personnel is involved.

3.6.3. Absorption of Radiation, Dose Rate Calculations, and Shielding

For practical purposes we may assume that the radioisotopes most likely to be used in analytical applications emit negatrons (β^-) and possibly γ rays. The following discussion is limited to such isotopes.

Negatrons (β^-) lose their energy by interaction with electrons of the atoms in the material into which they pass. Clearly, therefore, from the viewpoint of shielding, the more electrons per unit surface area in their path, the better. Although the density of the absorbing medium has some effect, the 'surface density' or weight per unit area is more important. Calculations show that materials with low atomic number are particularly effective, and for this reason use is frequently made of *aluminium*, or materials such' as *high density polythene*.

γ rays lose their energy in a very complex manner, which depends very much on their initial energy. However, it can be shown that there is an exponential relationship between the number of γ rays (N_0) incident on an absorber of thickness x, and the number emerging (N)

$$N = N_0 \, e^{-\mu x}$$

where μ is the 'absorption coefficient' of the absorbing material concerned.

In the same way that in Section 1.3 the 'decay constant' for describing radioactive decay rates can be more practically replaced by the idea of 'half-life', so in this situation the absorption coefficient is replaced by the use of 'half-thickness', the thickness of a given absorber that will reduce the γ dose rate to half the initial value.

Thus the half-thickness of a given absorber would reduce an exposure dose rate of for example 2×10^{-3} to 1×10^{-3} roentgen h^{-1}.

Half-thickness values are available for a variety of materials, and enable calculations to be made of suitable thicknesses of shielding for sources of a given strength. However, such theoretical values should be treated with caution; practical factors such as the geometry of the source, the geometry of the shielding, and scattering of the radiation must be considered. The most effective materials for shielding γ rays have a *high density* eg lead. However, lead is expensive, and in some situations less expensive materials such as concrete or water may be more appropriate.

SAQ 3.6a	The dose rate at the surface of a ^{60}Co γ ray source is 160 mrem h^{-1}. Given that the half-thickness value for lead is 1.25 cm, what thickness of lead shielding is necesssary to reduce the dose rate to 10 mrem h^{-1}? $\longrightarrow$

SAQ 3.6a

SAQ 3.6b   Given that for ^{60}Co the absorption coefficient of concrete is $\mu = 0.075$ cm^{-1}, calculate the thickness of concrete necessary to achieve the reduction of dose rate from 160 to 10 mrem h^{-1}.

Your answers to the preceding SAQ's 3.6a and 3.6b suggest roughly seven times the thickness of concrete is required compared to lead. This is a relatively crude estimate, but is put in perspective when you realise that on a volume basis lead is several hundred times more expensive than concrete.

Before we can work out the amount of shielding needed for a given source we need to be able to calculate the dose rate it is producing. Again, the basis of the various equations is complex, and they are given below principally to give you a 'feel' for the kind of dose rates that might be experienced in typical laboratory situations.

(*a*) Gamma Radiation

For a 'point source' (ie one for which the volume and density are insignificant compared with its radioactive content):

In pre SI units:

$$\text{dose rate at 1 metre} = 0.53 \; A \; E \; \text{rad h}^{-1} \qquad (3.6a)$$

where A/Ci is the activity and E/MeV is the *total* energy per disintegration.

In SI units:

$$\text{dose rate at 1 metre} = 143 \; B \; E \; \text{Gy h}^{-1} \qquad (3.6b)$$

where B = activity in GBq (1 GBq = 10^9 Bq).

(*b*) Beta Radiation

In this case, because of the different mechanisms of energy loss, the dose rate is almost independent of energy. The working approximations for a point source are:

In pre SI units:

$$\text{dose rate at 10 cm} = 3000 \; A \; \text{rad h}^{-1} \qquad (3.6c)$$

In SI units:

dose rate at 10 cm $= 0.81\ B$ Gy h^{-1} (3.6d)

SAQ 3.6c   Calculate the dose rate in SI units for the following 0.37 GBq sources and distances:

(*i*) a β^- source at 10 cm;

(*ii*) a γ source of total energy 2.5 MeV at 1 m.

In conclusion we note two points. Firstly, because β^- particles are relatively easily stopped by low atomic number elements, it is almost inevitable that except for β^- particles of the highest energy there will be considerable absorption of the β^- particles *within the source material itself* (self-absorption). This will be a major problem for β^- emitters of low energy, and will mean that much care will be needed in preparing sources for counting, and in making appropriate allowances for self absorption when interpreting observed count rates. Secondly, it is instructive to consider the effects of having *no shielding* (other than air). For highly penetrating γ rays the dose rate is inversely proportional to the square of the distance from the source

ie   dose rate $= k/d^2$

Thus if the dose rate is 20 μGy h^{-1} at 1.0 cm it will be 20/10^2 (ie 0.2 μGy h^{-1}) at 10 cm, and 0.2 $\times$ 10^{-2} μGy h^{-1} at 1 m. However, except at very short ranges, the inverse square law does not apply to β^- radiation.

In practice this means two things:

(i) distance can be a very effective shield; depending on the manipulation necessary, it may well be better in handling a gamma emitter to use metre length tongs rather than use lead brick shielding, and will certainly be less expensive;

(ii) at *short* distances the intensity of the effect from β^- emitters will be greater than from gamma emitters of the same initial energy.

To convince yourself of this try the following problem.

∏   Compare the dose rate for a pure β^- emitter (^{32}P) with that for a γ emitter, both of activity 10 mCi and $E = 1.7$ MeV. What will the dose rate be at 10 cm for the γ emitter?

For the β radiation (using equation 3.6c) the dose rate at 10 cm

$= 3000 \times 10 \times 10^{-3} = 30$ rad h^{-1}

For the γ radiation (using equation 3.6a):

the dose rate at 1 metre

$$= 0.53 \times 10 \times 10^{-3} \times 1.7 = 9 \times 10^{-3} \text{ rad h}^{-1}$$

At 10 cm the dose rate for the γ emitter will be 10^2 greater (according to the inverse square law), ie 9×10^{-1} rad h^{-1}.

The moral is that whereas the whole body exposure from a β^- source may be negligible if it is manipulated at arms' length, the dose to the fingers may be quite significant, and this will be important when carrying out dilutions from concentrated solutions of radioisotopes.

SAQ 3.6d

In SAQ 3.6c part (*ii*), you calculated the γ dose rate at 1 metre. Will the γ dose rate at 10 cm be greater or lesser than that for the β emitter you calculated in part (*i*)?

3.6.4. Radiation Monitoring in a Laboratory Situation

If you become involved in radiation work you will find that laboratories are *classified* according to the nature of the work to be undertaken in them. Factors such as use of open or closed sources, strength of sources, and toxicity of the isotopes have to be considered. The facilities necessary for the various types of work vary, not just in design or layout but in construction materials, services, ventilation etc. As you have seen in the previous sections it will be necessary to have available all the necessary shielding and handling facilities. It will also be necessary (though we shall not discuss it here) to make arrangements for planned and regular contamination checks, decontamination procedures, and routes for disposal of material after use.

Broadly speaking we can distinguish between personnel monitoring and laboratory monitoring.

Personnel Dosimetry

There are two possible ways in which the dose to an individual can be recorded; one system enables the dose at any given time to be checked, whilst the second simply records retrospectively the total dose over a given working period.

The first type is a very simple quartz fibre electroscope, shaped very much like a fountain pen, to fit into the pocket of a laboratory coat. The principle is very simple; the electroscope is charged so that the fibre moves away from its support. As radiation enters the electroscope it causes ionisation, which discharges the electroscope and causes the fibre to return towards the support (Fig. 3.6a).

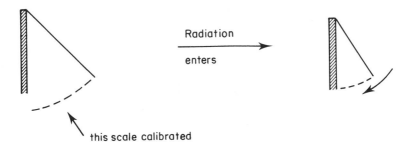

Fig. 3.6a. *Effect of radiation on an electroscope fibre*

The scale of deflection is directly calibrated in units of dose; by holding the scale towards a source of illumination the worker can regularly check his/her dose, which is a major advantage. The disadvantage is that the need for reasonably robust construction limits such devices essentially to work with γ emitters.

The retrospective devices are of two kinds: film badges and (more recently) thermoluminescent phosphor detectors. The radiation received causes a darkening of the photographic film, or an excitation of the phosphor, respectively. In either case the intensity of the effect is directly related to the amount of radiation received.

The advantage of such devices is that they enable a permanent record to be kept for every radiation worker; the disadvantage is that the dose received is not known until *after* the device (photographic film or phosphor) has been treated.

Finally you should note that after carrying out radiation work and before leaving the laboratory it is standard practice for personnel to monitor themselves and their clothing for contamination.

Laboratory Monitoring

It is a legal requirement that whilst work is in progress with radioactive materials, monitoring facilities must be available and functional. There is a wide variety of such equipment available, and the choice of monitor very much depends on the type of experiment and the nature of the isotope. The detector systems cover the whole range we have discussed earlier, Geiger counters, scintillation counters, solid-state detectors, and ionisation chambers. Monitors may be either portable for use at the working site, or in fixed positions around the laboratory. Many versions give not only a visual display of counting rate (either directly in disintegrations per unit time or in dose rate for selected types of radiation) but also an audible response. In science fiction Geiger counters always have an audible response; in practice, for monitoring purposes it is an optional (though very useful) extra!

An important aspect of laboratory monitoring is the need for frequent checking and calibration of monitors, and standard reference sources are commercially available for this purpose (see 2.1.3).

SUMMARY AND OBJECTIVES

Summary

Radiation protection is subject to rigorous legal controls. Dose rates can be expressed in various ways depending upon the type of radiation being considered and the circumstances of exposure to, and/or absorption of, radiation. For similar reasons shielding materials are chosen to match the nature of the source. Since the dose rate for γ radiation decreases with the square of the inverse of distance, and since β^- radiation travels relatively short distances, it is common practice to use distance as well as shielding for protection.

Objectives

You should now be able to:

- recognise the different ways in which individuals may be exposed to radiation;

- understand the different systems for expressing dose rate;

- calculate dose rates for given situations;

- evaluate any necessary protective measures, particularly in relation to shielding;

- identify the most common systems for personnel and laboratory monitoring.

4. Radioanalytical Methods

4.1. METHODS OF DIRECT DETERMINATION

Overview

The section specifies elements for which measurement of natural radioactivity is a possible method of determination. It distinguishes between elements which have one or more naturally occurring radioisotopes and those for which all radioisotopes are man-made. The section gives a quantitative example of the determination of potassium, and the most important radiometric techniques for elements with atomic number >83.

All isotopes of elements with atomic number >83 (ie above Bi in the Periodic Table) are radioactive. Frequently such elements occur together, either as a result of being in the same natural decay chain (Section 1.2) or as products of the capture (without fission) of neutrons by uranium. Also, a few of the elements with atomic number <83 have a naturally occurring radioactive isotope (eg ^{14}C, ^{40}K, ^{87}Rb). Provided that it is possible to count such isotopes without interference from other radioisotopes present, it should be possible to relate the measured activity to the concentration of the element concerned.

4.1.1. Elements with Atomic Number Less Than or Equal to 83

Probably the best example is the determination of potassium. Naturally occurring potassium contains 0.012% of ^{40}K, for which $t_{0.5} = 1.28 \times 10^9$ years and which decays by negatron and gamma emission. The specific activity is 1.85×10^3 disintegrations $min^{-1} g^{-1}$ of potassium; although this is low it is possible using rather concentrated solutions, to obtain a calibration curve of activity versus concentration. Thus by measurement of the activity of solutions of unknown potassium concentration it should be possible, by referring to the calibration curve, to calculate the concentration of the unknown.

This sounds straightforward; what are the advantages and problems? The advantages are that the method is quite rapid, and, in the absence of other radioisotopes, is free from interferences. The problems derive from the low specific activity, which normally leads to a requirement for pre-concentration and the counting of rather dense solutions; this in turn leads to the need for corrections due to self-absorption (ie absorption of radiation in the source before it can enter the counter). Because of these problems the method is not in widespread use.

SAQ 4.1a	Assuming that a liquid Geiger counter has an efficiency of 30% for ^{40}K radiation, calculate the expected count rate for a 10 cm^3 sample of KCl solution containing 50 g dm^{-3} KCl. [$A_r(K) = 39.1$, $A_r(Cl) = 35.45$, specific activity $= 1.85 \times 10^3$ disintegrations $min^{-1} g^{-1}$ potassium].

The answer you have obtained from the SAQ should reinforce the problem of low count rate: 10 cm^3 of a solution of 5% w/v KCl gives, theoretically, 485 disintegrations per minute. This is not a method for low limits of detection!

Other related methods of direct determination are of a more specialised nature—the so-called 'carbon-dating' technique effectively uses measurements of the ^{14}C content of natural specimens that have been removed from the natural carbon cycle, and there are a few methods for the analysis of rocks that hinge on measurements of natural radioactivity. One such is exemplified by SAQ 4.1b.

SAQ 4.1b
All naturally occurring rubidium ores contain ^{87}Sr resulting from β^- decay of ^{87}Rb. In naturally occurring rubidium, 278 of every 1000 rubidium atoms are ^{87}Rb. A mineral containing 0.85% rubidium was analysed and found to contain 0.0098% strontium. Assuming that all this strontium originated from decay of ^{87}Rb, estimate the age of the mineral.

($t_{0.5}$ of ^{87}Rb = 6.2 × 10^{10} years).

4.1.2. Elements with Atomic Number Greater Than 83

The most important analytical procedures using the radioactivity of isotopes of these elements are directly or indirectly attributable to the nuclear power programme, and particularly concern the transuranic elements which build up slowly in nuclear fuel:

$$^{238}_{92}U + ^{1}_{0}n \rightarrow ^{239}_{92}U$$

$$^{239}_{92}U \rightarrow ^{239}_{93}Np + ^{0}_{-1}\beta \qquad t_{0.5} = 23.5 \text{ min}$$

$$^{239}_{93}Np \rightarrow ^{239}_{94}Pu + ^{0}_{-1}\beta \qquad t_{0.5} = 2.355 \text{ days}$$

The products of this sequence, and particularly $^{239}_{94}Pu$ which has a half-life of 2.44×10^4 years, will also be able to capture neutrons to give other isotopes of these elements.

∏ Complete the sequence of reactions

$$^{240}_{94}Pu \longrightarrow ^{240}_{\square}X + ^{0}_{-1}\beta$$

$$^{240}_{\square}X + ^{1}_{0}n \longrightarrow ^{241}_{\square}Y$$

$$^{241}_{\square}Y \longrightarrow ^{241}_{\square}Z + ^{0}_{-1}\beta$$

By checking with a copy of the Periodic Table you can tell that the products are $^{240}_{\square}X = ^{240}_{95}Am$, $^{241}_{\square}Y = ^{241}_{95}Am$, $^{241}_{\square}Z = ^{241}_{96}Cm$

These isotopes decay by processes involving α particle emission, the α particles having characteristic energies as shown below for the three most important isotopes of plutonium.

Isotope	α energy/MeV	
^{239}Pu	5.143,	5.155
^{240}Pu	5.124,	5.168
^{241}Pu	4.897	

Hence, analysis of samples containing mixtures of such isotopes will require the ability to discriminate between the various α particles on the basis of their energies. This technique, known as α particle spectrometry, is now highly developed for the analysis of environmental samples (particularly effluents, marine samples, vegetation and foodstuffs). The processes are characterised by very careful control of the sampling, recovery sequence and preparation of the sample for counting.

SUMMARY AND OBJECTIVES

For radioactive isotopes of naturally occurring elements, and for those elements for which all isotopes are radioactive, it is possible to set up analytical methods for relating the amount of activity measured to the amount of the element present. The principal experimental restriction is that other radioactive isotopes should either be absent, or should not interfere in the counting procedure used.

Objectives

You should now be able to:

- recognise the elements for which determination by measurement of natural radioactivity is possible;

- appreciate the problems in such determinations;

- explain the significance of α emission for the transuranic and related elements;

- relate natural levels of activity to possible count rates.

4.2. TRACER INVESTIGATIONS OF EXISTING ANALYTICAL METHODS

Overview

This very brief section shows that tracer investigations of methods used in analytical chemistry do not differ in principle from tracer applications in other areas of chemistry. It gives examples of typical procedures and points out that they are principally in the area of separation techniques.

The principles underlying this application to analytical processes are identical with those of any tracer application in chemistry. Because very small quantities of a radioisotope will normally give a high count rate only small samples need be added to the system being studied, thus giving no interference with the system. Further, a molecule labelled with a radioisotope will follow the identical chemistry as unlabelled molecules. Only one radioisotope will normally be used, so that sampling and counting can be straightforward.

4.2.1. Typical Procedures

Most of the processes which have been studied involve separation and/or concentration of analytes (ie the species being measured). Examples include liquid–liquid extraction processes and ion-exchange separations. We shall consider first the application to a study of the efficiency of gravimetric procedures.

Suppose we wish to measure the efficiency of precipitation of phosphate by a newly published method (or indeed an established method). Orthophosphate labelled with ^{32}P is readily available at a high specific activity, and can be diluted so that the amount of active phosphate added to the inactive sample is at an appropriate level for counting. The mixed phosphate solution is taken through the gravimetric procedure, and both precipitate and filtrate are collected. The filtrate is made up to a known volume, and since it is necessary to use the same counting method for filtrate and precipitate, the latter is redissolved and made up to a known volume.

Appropriate aliquots of the two solutions are counted in a suitable apparatus; in this example 10 cm^3 samples would be counted in a liquid Geiger counter.

The efficiency of the precipitation may be simply calculated as the ratio of the count rate in the precipitate to the combined count rate of precipitate and supernatant, normally expressed as a percentage. It should be evident that the major advantage, apart from the simplicity of the procedure, is that it is not necessary to dry and weigh the precipitate. Quite analogous procedures can be followed for the study of other analytical separations.

SAQ 4.2a	The filtrate from a gravimetric procedure for phosphate, labelled with ^{32}P was made up to a volume of 100 cm^3. An aliquot of 10 cm^3 gave 1750 counts when counted for 5 minutes in a liquid Geiger counter. The precipitate from the procedure was dissolved in exactly 10 cm^3 of solution, and gave 14 500 counts min^{-1} when counted under identical conditions. The efficiency of precipitation is:

A   97.64%

B   80.55%

C   45.31%

D   89.23%

The detail of chromatographic separations is covered elsewhere in this course. However in this section we should draw attention to the great influence that radioactive tracers have had upon ion-exchange separations in particular. From your previous studies in chemistry you may remember that the lanthanoid elements (previously called the lanthanide elements) are chemically very similar, and until some 30 years ago were considered very difficult to separate. [If you have not studied these elements before, look at a Periodic Table; they are the elements from lanthanum ($Z = 57$) to lutetium ($Z = 71$)]. Work on nuclear fission of uranium revealed two important things in relation to the lanthanoids:

(a) many radioactive isotopes of the lanthanoids are fission products, and

(b) the transuranic elements such as Np, Pu, Am (see 4.1.2) are chemically very similar to the lanthanoids (they are often called actinoids).

This stimulated the search for new analytical and separation methods for these elements, culminating in the classical ion exchange separations involving adsorption of the ions on ion exchange resins and their subsequent elution with complexing agents. The identification of the individual lanthanoids and actinoids is made relatively easy by measuring the activity in the fractions as they are eluted from the ion exchange column. Typical elution data, showing the removal of each element from the column as a function of volume of eluant (and therefore time), are in Fig. 4.2a.

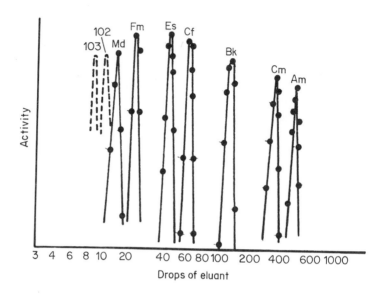

FIG. 4.2a

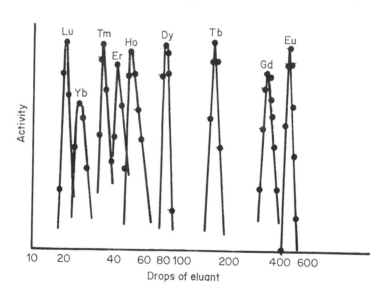

Fig. 4.2a. *Elution of tripositive lanthanoid and actinoid ions from Dowex-50 ion-exchange resin with ammonium α-hydroxyisobutyrate. The predicted positions for elements 102 and 103 are indicated by dotted lines. (After G. T. Seaborg,* Man-made Transuranium Elements, *Prentice Hall 1963.)*

This whole area of work, for part of which Seaborg and McMillan received a Nobel Prize in 1951, although not strictly a part of this Programme, is a fascinating topic in the modern history of chemistry.

As you are probably aware, the technique of liquid–liquid extraction (solvent extraction) is widely used in analytical chemistry. Radioactive tracers are an excellent means of studying the partition of compounds between two immiscible phases, and the efficiencies of many such processes have been measured in this way. Parameters such as distribution ratios and partition coefficients are readily calculated, and in some cases the method can be extended to allow calculation of the formation constants of the metal complexes on which many solvent extraction separations depend.

SAQ 4.2b

In the nuclear industry, uranium in the form of the UO_2^{2+} ion is extracted from an aqueous to an organic phase using the complexing agent tri-n-butylphosphate (TBP) dissolved in kerosene. In order to carry out a laboratory experiment to investigate the efficiency of this separation an aqueous solution containing $UO_2(NO_3)_2$ was shaken with an equal volume of TBP/kerosene. After the phases had been separated 10 cm^3 of each phase was counted in a liquid Geiger counter. The organic phase gave 11 300 counts min^{-1} and the aqueous phase gave 4394 counts min^{-1}.

Calculate:

(*i*) the % efficiency of the separation;

(*ii*) the number of extractions necessary to extract >99% of the uranium into the organic phase. $\longrightarrow$

SAQ 4.2b

SUMMARY AND OBJECTIVES

Summary

Because of the ease of detection of minute quantities of radioactive isotopes, they are well suited as tracers for investigation of analytical processes depending on concentration and/or separation. Particular areas of application are solubility studies, solvent extraction, and ion-exchange processes.

Objectives

You should now be able to:

● appreciate the principles underlying the use of radioisotopes as tracers for analytical separations;

● be familiar with typical applications of this type;

● calculate separation efficiencies using typical data.

4.3. ANALYTICAL METHODS INVOLVING DIRECT ADDITION OF RADIOISOTOPES TO THE SYSTEM

Overview

The section describes the four most widely used forms of isotope dilution analysis, and indicates the important features of each. It is shown that the most important techniques are direct isotope dilution analysis, and the version known as radioimmunoassay, and these methods are described in some detail.

There are two ways in which radioactivity may be introduced into a system as a means of analysing for one of the components. The most obvious way is through the addition of a tracer, followed by some quantitative study of its distribution in the system; such methods are the various forms of isotope dilution analysis. The other way is the irradiation of the sample in such a way that the element to be determined becomes radioactive, and the induced activity is quantitatively measured independently of any other activity that may be induced; this is the basis of activation analysis, and is considered in Section 4.4.

4.3.1. Isotope Dilution Analysis (IDA)

We are probably all familiar with the classical analytical problem: 'this is a mixture of X, Y, and Z; please measure the amount of X that is present'.

In order to achieve this by IDA we shall need:

(*a*) a radioactive form of X *of known specific activity*;

(*b*) a satisfactory method of counting the radioactive form of X; and

(*c*) a separation procedure which will give a pure sample of X.

Probably the best way to understand IDA is to follow the steps in a typical procedure.

Take a known aliquot of the mixture containing X (assume the known amount of X is w_x), and add to it a known aliquot of a radioactive form of X of known specific activity, $S_O = A_O/w_O$. Allow the mixture to equilibrate. Separate a *pure* sample of X, and redetermine the specific activity; The extent to which the specific activity has been diluted will be a measure of the amount of unknown X originally present.

The arithmetic for such a procedure is quite simple. The new (diluted) specific activity of the sample is given by:

$$S_1 = A_O/(w_O + w_x)$$

(Note that the total activity in the system is still A_O)

$$\therefore \quad (w_O + w_x) = A_O/S_1$$

On rearrangement this becomes:

$$w_x = (A_O/S_1) - w_O$$

Since $S_O = A_O/w_O$ we have $A_O = w_O S_O$

and hence $w_x = w_O S_O/S_1 - w_O$

$$\therefore \quad w_x = w_O[(S_O/S_1) - 1]$$

Now, w_O and S_O are known, and S_1 is being measured. Hence we can easily calculate w_x. Before we do this, let us consider a typical determination and particularly look at the levels of radioactivity necessary.

A good example is the determination of the amino acid glycine, NH_2CH_2COOH, in a mixture of other amino acids. Our first requirement is for radioactive glycine of known specific activity. It is available at a specific activity of 1.85 GBq mmol^{-1} as ^{14}C labelled glycine.

Π   You have previously carried out conversions of specific activity from the units given here to units of disintegrations min^{-1} mg^{-1}. Try that calculation again for this sample (If you cannot remember how to do it, look back to SAQ 1.4b).

Your answer should be 1.48 × 10^{12} disintegrations min^{-1} mg^{-1} which of course is far too high to handle without dilution—a typical tracer experiment might involve count rates of 10^3–10^4 disintegrations minute^{-1}.

We can now consider some typical figures.

The labelled glycine as purchased is diluted with pure unlabelled glycine so that the specific activity (allowing for the efficiency of the counter) is 18 000 counts min^{-1} mg^{-1}. 5 mg of this diluted labelled glycine is now added to a sample of the amino acid mixture, and after equilibration a small quantity (about 1 mg) of pure glycine is isolated. Chromatographic methods are often used for this separation.

Note that the amount of pure glycine isolated is much less than was originally added. As you will see later, this does not matter provided that the sample isolated is pure.

The specific activity of the sample isolated is determined and found to be 1500 counts min^{-1} mg^{-1}.

∴ $S_1 = 1500$ counts min^{-1} mg^{-1}

Hence from Eq. 4.3a

$w_x = 5.0\,[(18000/1500) - 1]$ mg

$ = 5.0\,(12.0 - 1\,)$ mg

$ = 55$ mg

The calculation could hardly be more simple, nor—in principle—could the method. So let us look again at the practical requirements:

>   (*a*) the species to be determined must be available in a radioactive form of known specific activity and high purity;

>   (*b*) it should mix fully with the inactive form in the unknown; and

>   (*c*) there should be a separation method capable of giving a pure sample of the species to be determined.

Now let us return to the recovery of pure material (in this case glycine). We have assumed that we only isolated 1 mg of pure glycine, whereas we know that we originally added 5 mg and (as it turns out) there was 55 mg of inactive glycine also present.

So, it would seem that we do not need a quantitative separation or 100% recovery of unknown. You can prove this by completing the calculation.

∏   Assuming that only 10% of the total (X + labelled X) is recovered calculate the new specific activity.

Only 10% of the activity will have been recovered. Similarly only 10% of the total weight will have been recovered. Hence the same recovery factor will occur in both the numerator and denominator of the equation for specific activity:

$$S_1 = 0.1\, A_0/0.1\, (w_0 + w_x)$$

$$\therefore \quad S_1 = A_0/(w_0 + w_x)$$

Yes! This is the same as if we had isolated all of X. ie S_1 is independent of yield of (diluted, radioactive) X.

Thus, the great value of IDA is that, provided a *pure* sample of X can be obtained for determination of S_1, *quite wasteful* and often *rapid* separation procedures can be used. This is often very useful in the isolation of a particular component of a mixture of similar compounds, eg a chromatographic separation of components of a mixture from a column need not be quantitative (Fig. 4.3a).

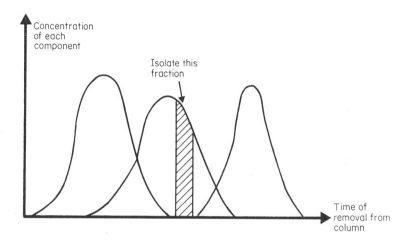

Fig. 4.3a. *Potential use of a non-quantitative separation*

The most common separation techniques are solvent extraction, ion exchange, electroplating, and simple precipitation of derivatives of X.

SAQ 4.3a

It is required to measure the orthophosphate content of a solution X. A 1 cm^3 sample of X, of density 1 g cm^{-3}, was taken, and 3.0 mg of ^{32}P-labelled PO_4^{3-} of specific activity 0.09 μCi mg^{-1} was added. From this mixture a pure sample of orthophosphate was isolated; the sample weighed 30 mg, and was found to give 1.8 × 10^4 counts min^{-1} in a counter of 30% detection efficiency. Calculate the concentration of orthophosphate in X, expressing your answer as % w/w. $\longrightarrow$

SAQ 4.3a

So, IDA has many attractions. Are there serious problems? In practice the only one of real significance is the requirement to measure very small amounts of labelled, diluted X. This latter point is potentially a limiting factor; it may not be possible to measure a weight of an isolated sample of X that matches the potential sensitivity of the measurement of the radioactivity of that sample.

Let us think back to our previous data. We assumed that our isolated diluted sample weighed 1 mg and gave 1500 counts min^{-1}. Measuring 1500 counts min^{-1} accurately is much more easy than measuring 1 mg accurately!

4.3.2. Substoichiometric Isotope Dilution Analysis

This technique was developed to try to overcome the problem noted at the end of the previous paragraph for 'conventional' IDA ('Substoichiometric' means 'less than the completely equivalent amount'). The method aims to isolate *equal but substoichiometric amounts* of a derivative of the analyte X, by a method known to give quantitative results. Suppose it is desired to measure (say) w_x μg of SO_4^{2-} in a mixture. The procedure might be:

(*a*) add a known amount (w_O) of SO_4^{2-} labelled with ^{35}S to a known aliquot of the unknown in solution;

(*b*) equilibrate;

(*c*) add a *known* amount of Ba^{2+} which is *just less than* equivalent to w_O. This causes precipitation of *some* SO_4^{2-} as $BaSO_4$;

(*d*) filter the precipitate. *NB*: There is no need to weigh it;

(*e*) measure the activity of the precipitate;

(*f*) repeat (*a*)–(*e*) but without any unknown.

Again there is no need to weigh the $BaSO_4$.

Why?

The same amount of precipitate will be obtained in each case because the reaction of Ba^{2+} with SO_4^{2-} is quantitative, and the added Ba^{2+} is less than equivalent to w_O. The limiting step is thus the accuracy of measurement of activity.

There have been several excellent recent examples of this technique; a good example is the paper by R. A. Pacer and S. M. Benecke, *Analyt. Chem.* (1981), **53**, 1160–3, which is considered further in 5.1.1.

4.3.3. Derivative Dilution Analysis

This is a version of IDA in which the radioisotope is added to the system in a *reagent* which reacts reproducibly with the species being determined to give a *derivative* which can be separated and counted. The principles are not significantly different from IDA and for simple chemical examples the method is not of great importance. However, it has led to a method of great importance in analytical biochemistry, known as radioimmunoassay, and although the detail of this technique is dealt with in another part of this course, the basic principles are discussed here.

4.3.4. Radioimmunoassay (RIA)

The most important development of IDA is the technique known as radioimmunoassay, which competes with neutron activation analysis (Section 4.4) as the most widely used radioanalytical method. It derives from a version of IDA, known as *derivative dilution analysis*, in which the radioisotope is added to the system in a *reagent* which reacts reproducibly with the species being determined, to give a *derivative* which can be separated and counted. Its development into RIA came with the recognition that many reactions of biochemical or metabolic importance, which take place *in vivo*, involve derivatisation reactions which might be reproduced *in vitro*.

Biological fluids may contain materials (often called *immunogens* or *antigens*) which may lead to some metabolic malfunction and/or illness. The body may react to produce materials (known as antibodies) which will counteract the immunogens. In order to study the course of *in vivo* immunogen–antibody reactions (and hence of illness or metabolic disorder resulting from the immunogen) it may be necessary to measure trace quantities of immunogens in biological fluids taken from a patient. If the *in vivo* reaction can be carried out reproducibly (and preferably quantitatively) *in vitro*, it should be possible to apply radioisotopes in a derivative dilution analysis mode. The fundamental requirements are:

(*a*) to make a labelled form of the immunogen;

(*b*) to make an antibody (in this context usually called a *binding agent*) specific for the immunogen;

(*c*) to carry out a reproducible reaction between immunogen and antibody;

(*d*) to separate the immunogen–antibody 'complex', and determine the relative amounts of 'natural' immunogen and added (labelled) immunogen in the 'complex'.

Before the method is discussed in more detail it is necessary to note that although RIA was initially developed to measure levels of naturally occurring immunogens, it is now very widely applied to study the effects of externally administered compounds having metabolic results (eg drugs, steroids).

RIA may be carried out in several ways, but is best understood by considering the most common experimental procedure:

(*a*) to a known volume of the solution containing the unknown quantity of immunogen add a known volume of solution containing labelled immunogen of known specific activity;

(*b*) add a known quantity of binding agent;

(*c*) ensure complete equilibration; and

(*d*) separate the immunogen-binding agent 'complex', and measure the amount of activity remaining in solution relative to that bound in the 'complex' $(R^*_{f/b})$.

A simple pictorial representation, showing unknown immunogen as O, labelled immunogen as X, and binding agent molecules each having four sites of attachment is shown in Fig. 4.3b

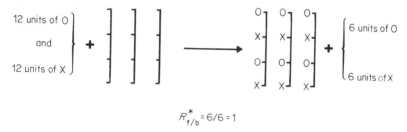

$$R^*_{f/b} = 6/6 = 1$$

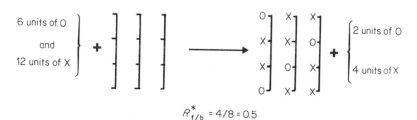

$$R^*_{f/b} = 4/8 = 0.5$$

Fig. 4.3b. *Pictorial representation of binding of immunogen molecules*

Clearly, $R^*_{f/b}$ is directly related to the amount of unknown immuno-gen originally present, and by using a simple calibration procedure it should be possible to determine the concentration of such un-knowns.

SAQ 4.3b

In order to set up an RIA procedure for measuring insulin a calibration curve was first prepared. Standard volumes of insulin solution of known concentration were mixed with a fixed volume of labelled insulin, such that the final concentration of insulin in each sample was 3, 5, 7 and 9 ng cm^{-3} respectively. The total activity of each solution was 2×10^4 counts min^{-1}. The same quantity of antibody was added to each solution, the mixture equilibrated, and in each case the insulin-antibody complex was isolated and the activity measured. The same procedure was then followed using a known volume of a solution containing an unknown amount of insulin.

The results are tabulated.

Concentration of insulin/ng cm^{-3}	3.0	5.0	7.0	9.0	Unknown
Activity of bound complex/counts min^{-1}	13 245	11 111	9 852	9 091	10 100

Draw an appropriate calibration curve, and hence calculate the concentration of insulin/ng cm^{-3} in the 'unknown' solution for which the count was taken. ⟶

SAQ 4.3b

The apparent simplicity of the method when shown schematically as above belies several practical problems. These are :

(*a*) the method requires highly pure antibody and (labelled) immunogen, and the synthesis of these is frequently very demanding;

(*b*) it is necessary to use a radioisotope which not only has acceptable characteristics in terms of decay scheme and ease of counting, but which also can be incorporated into an immunogen without undue difficulty. The radioisotopes most frequently used are:

$$^{125}\text{I} : t_{0.5} = 60 \text{ days}, \qquad E_\gamma = 0.035 \text{ MeV}$$

$$^{3}\text{H} : t_{0.5} = 12.26 \text{ years}, E_\beta = 0.015 \text{ MeV}$$

The iodine isotope is used because it can be inserted into appropriate organic derivatives, such as the tyrosine residue, on treatment with chloramine-T and NaI^*.

Chloramine-T is a readily available mild oxidising agent.

An example of the complexity of molecules which can now be purchased in a labelled form is the steroid aldosterone-3-(O-carboxymethyl)oximino-(2-[^{125}I] iodohistamine):

^{3}H (tritium) can also be incorporated into organic molecules, frequently by controlled isotopic exchange.

Further problems of a practical nature are:

(*c*) the need to have good standard materials;

(*d*) the immunogen–antibody 'complex' should be easily separable from the reaction solution. A common method is adsorption on dextran-coated charcoal;

(*e*) the need for frequent recalibration.

Notwithstanding these quite demanding practical requirements RIA is a technique which has revolutionised many bioanalytical determinations principally because it is possible to analyse routinely for very low levels of very complex organic molecules using very small initial sample volumes.

Finally, it should be noted that both of the isotopes most commonly used emit radiation of low energy. This infers the use of liquid scintillation counting (Section 3.2); this is a specialised technique, the detail of which is not reproduced here.

Because of the predominantly clinical and bioanalytical applications of RIA most papers are published in specialist journals. However, an excellent and accessible recent paper concerns RIA for barbiturates in blood and urine, and this is discussed further in 5.1.1.

SUMMARY AND OBJECTIVES

Summary

The addition of known amounts of a labelled compound of known specific activity to a mixture containing that compound, followed by a redetermination of the diluted specific activity, allows the calculation of the amount of the particular compound in the mixture. This technique has been developed in several ways to meet specialised circumstances, and in the form of radioimmunoassay it has been particularly valuable in clinical chemistry.

Objectives

You should now be able to:

- appreciate the full range of dilution analysis methods;

- specify the particular features of individual techniques;

- identify the most important techniques;

- describe the advantages of individual techniques;

- explain the main areas of application of dilution analysis methods;

- treat typical data quantitatively.

4.4. ACTIVATION ANALYSIS METHODS

Overview

This section describes the basic theoretical principles of activation analysis, and illustrates the principles with particular reference to activation with neutrons. An expression for the build-up of activity is used to calculate typical levels of induced activity, and these are related to likely limits of detection. The typical steps in (neutron) activation analysis are described, and activation analysis other than with neutrons is briefly considered.

4.4.1. Introduction and Theoretical Background

In 2.1.1 you saw that when a sample containing inactive isotopes is suitably irradiated, some or all of the initially inactive isotopes may become radioactive. The amount of activity induced in a particular isotope is:

unique to that isotope, and

directly proportional to the amount of initially inactive parent isotope present before irradiation.

Without, for the time being, specifying the proportionality constant we could write:

Induced Activity = Constant × Weight of Isotope; $A = kw$ (4.4a)

(this is a simple and crude form of an equation we shall develop and use later).

How can we use the two observations above?

Firstly, we need to measure the induced activity for a particular isotope independently of any other induced activities. Let us suppose we irradiate a very simple mixture containing just two isotopes that are activated $A \rightarrow A^*$, $B \rightarrow B^*$. What we are saying is that we need to measure A^* and B^* independently; to measure $(A^* + B^*)$ will not be sufficient.

The simplest and traditional method is to separate A^* and B^* chemically after irradiation; the modern method is to use instrumental methods to count A^* and B^* independently, most commonly by γ ray spectrometry (Section 3.4). Before we go any further you may not be surprised to learn that this is an ideal situation since frequently far more than two isotopes are activated.

Secondly we shall either need to know the proportionality constant, or be able to eliminate it. As you will see shortly the proportionality constant is rather complicated, and so the best thing is to eliminate it. As an analyst, what would you do to eliminate problems of this kind? *You would check the method against a standard*, and this is exactly what is done here. A standard containing the isotope in which we are interested is irradiated and then counted under identical conditions to the sample containing the unknown amount of the isotope. So we shall have:

for the standard: $A_{standard} = kw_{standard}$

and for the unknown: $A_{unknown} = kw_{unknown}$

$\therefore \quad A_{standard}/A_{unknown} = w_{standard}/w_{unknown}$

We are measuring the two values of A, and we know $w_{standard}$ so we can calculate $w_{unknown}$. It could hardly be easier!

SAQ 4.4a A 1 g sample of an alloy containing a small quantity of gold was irradiated and counted under identical conditions to a standard containing 10 mg of gold. The unknown gave an activity of 500 disintegrations min^{-1}, and the standard gave 2000 disintegrations min^{-1}. Calculate the % (w/w) of gold in the alloy. $\longrightarrow$

SAQ 4.4a

Before we consider how this apparently simple method can be carried out experimentally we need to consider the theory in a little more detail. So far we have assumed irradiation relative to a standard. If we wish to calculate such things as limits of detection we really need the simple version of the equation 4.4a in a more clearly defined form. In particular we need to know more about the various factors contributing to the constant k, appearing in the final form of Eq. 4.4a.

So what are the variables that might affect the build-up of activity when a sample is irradiated? One obvious factor is *time*: the longer the irradiation time the higher the level of activity you might expect to be induced. But there is a complication; at any given time in the irradiation, some stable nuclei are being converted to radioactive ones whilst some of the radioactive nuclei previously formed are decaying back to stable ones.

Thus the overall build-up of activity is a balance between the activation reaction $A \rightarrow A^*$ and the radioactive decay rate of A^*.

∏   What was was the time parameter we used to help define the radioactive decay rate?

    It was the 'half-life', $t_{0.5}$. (If you can't remember, re-read 1.3.2.).

So it should not surprise you if $t_{0.5}$ for the product isotope is one of the terms in k which occurs in the fully developed form of Eq. 4.4a.

Quite frequently for experimental reasons it is necessary to have a delay between the end of irradiation and the start of counting, and this delay time (T) then needs to be included in the final form of Eq. 4.4a.

To summarise the time factors involved, we shall expect k to include the irradiation time t, the product half-life $t_{0.5}$, and the delay time T. It is important in calculations that these should all be in the same units.

The next factor that affects the build-up of activity is quite simple; you would probably guess that the more particles you use to bombard the target the greater will be the activity induced. Put in more scientific terms we would expect the flux of particles, which is given the Greek symbol phi (ϕ) to occur in the final form of Eq. 4.4a. For the most common technique, activation with neutrons in a nuclear reactor, the neutron flux is expressed (in SI units) as neutrons s^{-1} m^{-2}. Before the SI system was introduced the flux was expressed as neutrons s^{-1} cm^{-2} (ie a factor of 10^4 times smaller).

Finally we have to explore the ways in which we can include the point mentioned at the outset, that the amount of activity induced in a given isotope is unique to that isotope. It is best to take a simple example, so we shall consider the irradiation of chlorine. In its natural state the element chlorine has two stable isotopes: 75.77% is $^{35}_{17}Cl$ and the remaining 24.23% is $^{37}_{17}Cl$. When we irradiate natural chlorine we shall necessarily irradiate both of these isotopes; they will capture the irradiating particles (and become radioactive) but both to different extents. For example, if we irradiate with neutrons, and consider the simple reactions:

$$^{35}_{17}Cl + {}^{1}_{0}n \rightarrow {}^{36}_{17}Cl^*$$

$$^{37}_{17}Cl + {}^{1}_{0}n \rightarrow {}^{38}_{17}Cl^*$$

we find that the first reaction is roughly 100 times more probable than the second. The probability of every nuclear reaction such as those above is called the 'capture cross-section' and is given the Greek symbol sigma, (σ). For reasons that are beyond the scope of this treatment it has units of cross-sectional area. Traditionally this was given the unusual name of 'barns', where 1 barn $= 10^{-24}$ cm^2. On the SI system 1 barn $= 10^{-28}$ m^2. We shall certainly expect σ to occur in any version of Eq. 4.4a that we may develop.

Now let us think a little further about the effects of irradiating an *element*. We are saying that all the stable isotopes present will be activated to different extents. If we wish, we can express the activity induced in the element as a (rather complicated) sum of the activities induced in the individual *isotopes*. In practice it is much more easy to calculate the activities for the individual isotopes, and so we can expect to include in the final form of our equation some way of representing the proportion of the particular stable isotope with respect to the total number of stable isotopes present. In practice this is done by including the fractional abundance of the particular *isotope* of interest, f, and the atomic mass of the *element, A*.

Finally, there is an important point in relation to the units for the activity. We are interested in relating the induced activity to the mass of the isotope concerned, whereas so far we have been talking about activating the nuclei of individual atoms of the isotope. The relationship between atomic mass and number of atoms is the Avogadro constant ($N_A = 6.022 \times 10^{23}$ mol^{-1}), and we can expect this to appear in the final form of Eq. 4.4a.

Let us recap! The factors we shall expect to include are:

$$t, \ T, \ t_{0.5}, \ \phi, \ \sigma, \ f, \ A, \text{ and } N_A.$$

We shall not derive the equation; we shall simply quote it—you may feel that the introduction has been sufficiently complex! There are various forms of the equation, but for our purposes the preferred version relates the induced specific activity S to the various parameters we have discussed:

$$S = \frac{6.022 \times 10^{23}\sigma\phi f(0.5)^{T/t_{0.5}}(1 - 0.5^{t/t_{0.5}})}{(A/\text{g mol}^{-1})} \qquad (4.4b)$$

the units of S being disintegrations s^{-1} g^{-1}

In the SI system the standard unit of mass is the kilogram, and although for activation analysis we would not consider irradiating kilogram quantities the equation is often written:

$$S = \frac{6.022 \times 10^{26} \sigma \phi f (0.5)^{T/t_{0.5}} (1 - 0.5^{t/t_{0.5}})}{(A/\text{g mol}^{-1})} \qquad (4.4c)$$

the units of S now being disintegrations s^{-1} kg^{-1}

The equation looks rather fearsome! Because both the growth and decay of activity is exponential (Section 1.3) the equation can be expressed in a more obviously exponential form. If your mathematics is good enough you can recognise this by showing that:

$$(0.5)^{T/t_{0.5}} = \exp(-0.693 \ T/t_{0.5})$$

If you cannot do this, don't worry; we are only interested in how we can use the Eqs. 4.4b or 4.4c. The best way of doing this is to calculate a limit of detection for a typical analysis.

4.4.2.   Limits of Detection

We shall take a typical example for *neutron* activation analysis, since this is by far the most widely used version of the method. Let us consider the determination of manganese by the reaction:

$$^{55}_{25}\text{Mn} + ^{1}_{0}\text{n} \rightarrow ^{56}_{25}\text{Mn}$$

$^{55}_{25}\text{Mn}$ is the only naturally occuring isotope of manganese $[A_r(\text{Mn}) = 54.94]$ so $f = 1$. The product $^{56}_{25}\text{Mn}$ has a half-life of 2.58 h and the value of σ for this reaction is 13.3 barn, ie 13.3 $\times$ 10^{-28} m^2.

We now need to assume certain things about the experimental parameters of flux and time; these will be explained in the next paragraph. We shall assume that ϕ is 10^{16} neutrons s^{-1} m^{-2} (this is a typical value for a nuclear reactor), and that we irradiate for 10.32 h (ie 4 $\times$ $t_{0.5}$) and have a pre-counting delay time of 2.58 h (ie $t_{0.5}$). These time values are chosen a little arbitrarily to make the calculation easier.

Hence

$$S = \frac{6.022 \times 10^{26} \times 10^{16} \times 13.3 \times 10^{-28} \times 1 \times 0.5^1 \times (1 - 0.5^4)}{54.94}$$

The factor $0.5^4 = 0.0625$, so $(1 - 0.5^4) = 0.9375$

$$\therefore \quad S = \frac{6.022 \times 13.3 \times 0.5 \times 0.9375 \times 10^{14}}{54.94}$$

$$= 6.834 \times 10^{13} \text{ disintegrations s}^{-1} \text{ kg}^{-1}$$

This is the *theoretical* specific activity for $^{56}_{25}\text{Mn}$ produced under these conditions. We need to know what we might measure experimentally; as you have seen earlier in the section on detection methods radiation counters (in this case we might use a scintillation counter or a semiconductor device) are seldom 100% efficient. We shall simply assume in this case that our detector is 10% efficient.

$$\therefore \quad S = 6.834 \times 10^{12} \text{ counts s}^{-1} \text{ kg}^{-1}$$

It is customary to use counts min^{-1} rather than counts s^{-1} so we shall have:

$$S = 6.834 \times 60 \times 10^{12} \text{ counts min}^{-1} \text{ kg}^{-1}$$

$$= 4.100 \times 10^{14} \text{ counts min}^{-1} \text{ kg}^{-1}$$

Now we have seen (eg the calculation in 4.3.1.) that a convenient count rate in laboratory experiments is in the order of 10^3–10^4 counts per minute. To make our calculation simple let us assume we can easily measure 4.100×10^3 counts min^{-1} (in fact we could do better than this).

Hence if 1 kg is giving 4.1×10^{14} counts min^{-1} and we can easily measure 4.1×10^3 counts min^{-1} it follows that we can easily detect $4.1 \times 10^3 / 4.1 \times 10^{14} = 10^{-11}$ kg of manganese. Limits of detection are usually quoted in fractions of a gram, so in this case the value is 10^{-8} g of manganese. Do remember that this is for the experimental conditions we specified earlier.

∏   How could you improve on this limit of detection?

To answer this question, think about the factors in the equation that you can control.

Yes! These are the flux, the irradiation time and the delay time. By increasing the flux and the irradiation time, and by reducing the delay time you could improve the limit of detection.

SAQ 4.4b

Consider the determination of manganese by neutron activation analysis. The only naturally occurring isotope of manganese, $^{55}_{25}\text{Mn}$, has a half-life of 2.58 h, and a capture cross-section of 13.3 barn. Calculate the theoretical count rate (counts $\text{min}^{-1}\ \text{kg}^{-1}$) in a detector which has an efficiency of 10% under the following experimental conditions: neutron flux = 5×10^{17} neutrons $\text{s}^{-1}\ \text{m}^{-2}$; time of irradiation = $6 \times t_{0.5}$ and the precounting delay time = $0.5\ t_{0.5}$. Which of the following answers is correct?

(*i*) 1.076×10^{16};

(*ii*) 3.044×10^{16};

(*iii*) 2.899×10^{16};

(*iv*) 3.044×10^{17};

(*v*) 8.457×10^{12}; $\longrightarrow$

SAQ 4.4b

The answer you obtained from the self assessment question suggests a limit of detection under these new experimental conditions of 10^{-10} g.

The limits of detection we are suggesting here (10^{-8}–10^{-10} g), which could of course be improved with better detector efficiency, suggest that neutron activation is potentially a very sensitive analytical method. Manganese is a particularly favourable example because of the high value of σ and the convenient $t_{0.5}$ value for ^{56}Mn, but it is the case that neutron activation is a highly sensitive method for a large number of chemical elements.

We should conclude with one cautionary note. The preceding treatment has assumed that the determination of the activity induced in a given isotope is not subject to interference from the activity of other isotopes present. This freedom from interference is a major advantage of activation analysis, but if interferences do occur (which is relatively rare) the sensitivity is likely to be reduced.

4.4.3. Practical Aspects of Neutron Activation Analysis (NAA)

There are four main stages in any activation analysis procedure; it is convenient to describe them in the context of NAA since this is by far the most widely used method. The stages are:

Sample preparation, irradiation, post-irradiation treatment, and counting.

(a) Sample Preparation

The general requirements are the same as for any analytical method: the sample must be homogeneous and representative of the bulk of the material being assayed. The requirements that are more specific to NAA are those which relate to the irradiation stage. (You have learned about them in 2.1.1.)

∏   What are they?

The sample to be irradiated should be compact, involatile, in a suitable physical form (normally solid), and not subject to radiation damage.

If the sample does not conform to these requirements it must be pre-treated; for instance, many environmental samples such as large volumes of water will need to be concentrated and/or the analyte species precipitated. Typical techniques are ion exchange, solvent extraction, and electrodeposition.

(b) Irradiation

Broadly speaking there are two ways in which this can be achieved — by reactor irradiation, or by irradiation in a laboratory source. The former method was the original basis of NAA, and still is predominant. Although nuclear reactor technology is a very interesting topic it does not form part of this course. All that we need say is that reactors used for NAA are based on the fission reaction:

$$^{235}_{92}U + {}^1_0n \rightarrow \text{Fission products} + \text{(average) 2.5 neutrons} + \text{Energy}$$

The product neutrons are used to generate further fission reactions, in a controlled 'chain reaction'. For our purposes we need ask only two questions—what is the neutron flux in a typical reactor, and how readily available are reactors for NAA procedures?

The answer to the first question is that the flux can be controlled fairly accurately at a given level, which normally lies in the range 10^{14}–10^{18} neutrons s^{-1} m^{-2}. Access to reactors for NAA is normally on a commercial basis, and usually is not a problem. The sample and standard are irradiated at a predetermined flux for a given time, and then either transported to the analyst's own laboratory for treatment and counting, or counted at the reactor site.

Laboratory neutron sources fall into two categories: 'isotopic' sources, and neutron 'generators'. Until quite recently the former were all based on the nuclear reaction:

$$^{9}_{4}\text{Be} + ^{4}_{2}\text{He} \rightarrow ^{12}_{6}\text{C} + ^{1}_{0}\text{n}$$

A source of $^{4}_{2}\text{He}$ nuclei (ie α particles), typically an isotope such as $^{241}_{95}\text{Am}$ (americium-241), is mixed with beryllium. Unfortunately the flux is low; for $^{241}_{95}\text{Am/Be}$ sources a source of 1 curie ($3.7 \times 10^{10}\text{Bq}$) produces only 2.2×10^{6} neutrons s^{-1}. Note that this is neutron output, *not* neutron flux—there are no units of area.

SAQ 4.4c

> Assuming that a neutron source is sufficiently small to be considered as a 'point source', and by calculating the surface area of a sphere of 1 cm radius, calculate the neutron flux at 1 cm for a 3 Ci source of $^{241}\text{Am/Be}$.

Although the answer to SAQ 4.4c is only approximate it shows that the flux is many orders of magnitude less than that of a reactor, and hence such sources are little used in NAA: the sensitivity of most determinations is poor. However, recently another isotopic source has become available; this is the isotope $^{252}_{98}Cf$ (Cf = californium). This decays by spontaneous fission and emission of α particles; the fission reaction produces neutrons with a higher yield than that of $^{235}_{92}U$. The theoretical neutron output is 2.34×10^{12} neutrons s^{-1} g^{-1}, or roughly 4×10^9 neutrons s^{-1} Ci^{-1}.

This sounds tremendous, and potentially it is so. At present, however, ^{252}Cf is available only in small quantities (5 μg sources are typical), but as larger sources become available there is little doubt that the uses of this isotope for NAA will increase.

Neutron generators are based on a deceptively simple nuclear reaction, in this case between deuterium and tritium:

$$^2_1H + ^3_1H \rightarrow ^4_2He + ^1_0n$$

This reaction is not spontaneous, but can be induced by removing the electron from one of the atoms (usually ^2_1H) and accelerating the resulting charged particle (deuteron) through a high potential ($\approx$ 50 kV) to hit a target containing the other atom. The resulting neutrons are emitted with very high energy (14 MeV), which is 10 times greater than a typical reactor neutron, with major implications for usage and particularly shielding. Typical flux values are 10^{12}–10^{14} neutrons s^{-1} m^{-2}.

Unfortunately, there is a requirement for sophisticated electronic, vacuum, and sample supply systems, together with a need for considerable shielding; the resulting expense has been a limitation on usage. However, neutron generators have found some use in activation analysis for production of short lived isotopes of the lighter elements such as oxygen.

(c) Post-irradiation Treatment

This depends almost totally on the matrix which was irradiated, and on the activities induced. There are three main possibilities:

(*i*)   Only an isotope of the desired element is activated, in which case it can be counted without interference. This is ideal, but very rare.

(*ii*) The various activities produced are sufficiently different in the energy of the emitted γ rays to enable the desired isotopes to be counted independently. This has been greatly enhanced by the use of semiconductor detectors (see Section 3.3), and has led to neutron activation analysis becoming a completely instrumental method (INAA).

(*iii*) Even with semiconductor detectors it is not possible to resolve overlapping γ peaks. In this case a chemical separation will be necessary.

(d) Counting Step

Most neutron-induced reactions lead to γ emitting isotopes, which are more efficiently counted by scintillation detectors than by Geiger counters. Increasingly, the former are being superceded by semiconductor detectors. (Sections 3.1 to 3.3). However, as you may appreciate, counting of the induced activity is not enough; it is essential that the activity being counted derives only from the isotope of interest. Let us therefore return to two crucial points made earlier. Firstly, most products of neutron irradiation emit γ rays, and, secondly, the emitted γ rays have energies which are characteristic of the isotope emitting them.

Thus it is a requirement of the counting step that we can discriminate between the energies of the various γ rays and arrange to count them independently. This is the technique of γ *ray spectrometry*; the detail of the technique was given in Section 3.4, and we simply need to reiterate a few practical points:

(*i*)   the detector systems most used are NaI(Tl) crystals and Ge(Li) semiconductor devices;

(*ii*) the efficiency of Ge(Li) detectors is poorer than NaI(Tl), but the γ ray resolution is far better;

(*iii*) modern systems use multi-channel pulse height analysers;

(*iv*) computer treatment of γ ray spectra is now commonplace.

4.4.4. Influence of Computing

You may not be surprised to learn that as in many instrumental analytical techniques computers now hold a central position in NAA procedures, and particularly those which are totally instrumental (INAA). A typical flow chart is shown in Fig. 4.4a:

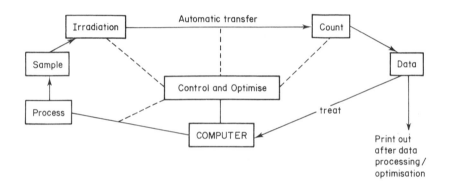

Fig. 4.4a. *Typical arrangement of instruments for INAA*

If you are particularly interested in this aspect you are recommended to read the paper by Lubkowitz *et al, Analyt. Chem.* (1980), **52**, 233–9, which contains a more detailed version of Fig. 4.4a.

4.4.5. Activation Analyses other than by Neutrons

These methods are primarily activation with γ rays (photon activation analysis) and with charged particles (CPAA). The former is little used, but the latter has been developed since 1970 as a method of some importance.

Experimentally it is now easy to accelerate (small) cations through a fairly high potential and direct them at a target containing the element to be determined. Typical examples are ^1_1H, ^2_1H and ^4_2He. The extent to which the charged particle must be energised (accelerated) to activate a chosen target nucleus can be calculated. There is normally no merit in increasing the energy significantly beyond this threshold value, since other side reactions may occur. Thus it is possible to determine boron by the reaction:

$$^{11}_{5}\text{B} + ^1_1\text{p} \ (3 \text{ MeV}) \rightarrow ^{11}_{6}\text{C} + ^1_0\text{n}$$

The product $^{11}_{6}\text{C}$, which is radioactive, is counted. At higher energies the threshold for other common isotopes is exceeded:

$$> 4.2 \text{ MeV} \qquad ^{14}_{7}\text{N} + ^1_1\text{p} \rightarrow ^{11}_{6}\text{C} + ^4_2\text{He}$$

$$> 19 \text{ MeV} \qquad ^{12}_{6}\text{C} + ^1_1\text{p} \rightarrow ^{11}_{6}\text{C} + ^2_1\text{H}$$

The biggest single problem is that penetration of charged particles into the matrix being activated is low. Thus, an early example claimed detection limits of 0.335 ppb on < 1 mg sample for the determination of oxygen in Cu by the reaction:

$$^{16}_{8}\text{O} + ^3_2\text{He} \ (7.5 \text{ MeV}) \rightarrow ^{18}_{9}\text{F} + ^1_1\text{H}$$

but at this energy penetration is roughly 25 μm.

Despite this disadvantage the method is valuable for some of the lighter elements which are not amenable to neutron activation analysis. An interesting recent example is the determination of fluorine in food.

Although not strictly an application of activation analysis, thin layer activation with charged particles is proving very advantageous in studies of the wear of engine components. Only the surface layer (or particular parts of that layer most susceptible to wear) are activated, and the wearing process is followed by measuring the appearance of activity in the lubricating oil.

SUMMARY AND OBJECTIVES

Summary

The activation of stable isotopes by irradiation of the target material with neutrons or charged particles is the basis of a highly sensitive analytical method for many elements. The activity induced in the desired element is measured, and since this is directly proportional to the amount of the element present, the latter can be calculated. It is now possible to carry out activation analysis on many samples completely instrumentally, thus avoiding any possibility of contamination and interference. It is common practice to compare the activity produced in the unknown to that produced in a standard, thus avoiding such problems as flux variation at the irradiation stage. Although neutron irradiation is most commonly carried out in a nuclear reactor, recent development of laboratory-based irradiation systems appear promising.

Objectives

You should now be able to:

● explain the theoretical background to activation analysis;

● specify the particular features of neutron activation analysis;

● calculate induced activity levels;

● relate induced levels to limits of detection;

● describe the important steps in a typical activation analysis;

● appreciate that activation other than by neutrons may be useful in certain situations.

5.   New Applications

5.1.   TYPICAL LITERATURE EXAMPLES OF
RADIOCHEMICAL METHODS IN ANALYSIS

Overview

By discussing and interpreting three published papers this section
aims to illustrate the importance of (neutron) activation analysis,
radioimmunoassay, and substoichiometric isotope dilution analysis.
The papers chosen are brief and are taken from readily available
analytical chemistry journals, rather than from the specialised radiochemistry journals.

This section should be read in conjunction with the relevant parts
of Section 4. One example is given for each of the three techniques
considered to be the most important in radioanalytical chemistry.
The examples have been chosen to give you an uncomplicated introduction to the ways in which the techniques concerned have been
used. Don't be put off by the fact that the applications may be in
areas with which you may not be familiar; what matters is that you
should see how the theoretical principles you have studied earlier
in the Unit are applied in practice.

5.1.1.   Activation Analysis

The paper with which you are provided (J T Tanner, M H Friedman,
and G E Holloway, *Analytica Chimica Acta* (1973), **66**, 456–9 is a
brief and relatively early example of neutron activation analysis, but
it illustrates the method almost ideally.

Perhaps you would now like to read through this paper.

456   *Analytica Chimica Acta*, 66 (1973) 456–459
             ⓒ Elsevier Scientific Publishing Company, Amsterdam – Printed in The Netherlands

SHORT COMMUNICATION

Arsenic and antimony in laundry aids by instrumental neutron activation analysis

JAMES T. TANNER and MELVIN H. FRIEDMAN

Division of Chemistry and Physics, Food and Drug Administration, Washington, D.C. 20204 (U.S.A.)

GERALD E. HOLLOWAY

Naval Research Laboratory, Nuclear Physics Division, Washington, D.C. 20390 (U.S.A.)

(Received 15th February 1973)

Considerable attention has been focused on work reported by Angino *et al.*[1], where the conclusion was drawn that arsenic from detergents (or more generally, laundry aids) are perhaps contaminating the rivers and drinking sources in Kansas and possibly other areas in the United States. As part of the Food and Drug Administration's heavy metal surveillance program, arsenic levels in laundry aids were measured. This paper describes the measurement of trace amounts of arsenic and antimony (which is also a toxic element) in laundry aids by neutron activation analysis (n.a.a.). The results for arsenic by n.a.a. are compared with those obtained by other analytical techniques. The concentrations in the various laundry aids tested ranged from 5 to 51 p.p.m. of arsenic and from 1 to 8 p.p.m. of antimony.

Experimental

Seven phosphate-based laundry aids (without boron) were purchased in the Kansas City District of the Food and Drug Administration. The samples included three enzyme presoaks, one detergent, two heavy-duty enzyme detergents, and one heavy-duty detergent, and were assumed to be representative of laundry aids obtainable elsewhere in the United States.

Samples of the laundry aids (*ca.* 500 mg) and standards of arsenic(III) oxide and metallic antimony were weighed directly into clean quartz vials, which were then sealed by using an oxygen–methane torch. Great care was taken to treat samples and standards alike and to avoid contamination of the quartz vials before neutron irradiation. After the samples and standards had been sealed in quartz, they were irradiated for 20 min in the high neutron flux position (*ca.* $1 \cdot 10^{13}$ n cm^{-2} s^{-1}) of the 1-MW reactor at the Naval Research Laboratory. The vials were rotated during the irradiation to insure a uniform neutron exposure. After the samples had been irradiated, they were allowed to undergo radioactive decay for 3–5 days to allow the 15.4-h ^{24}Na to decrease in intensity relative to the 2.7-day ^{122}Sb and 26.4-h ^{76}As. At the end of that time the samples were analyzed for arsenic and antimony.

The analyses were made with a high-resolution Nuclear Diodes Ge(Li) detector, which had a resolution of 2.09 keV at 1.33 MeV and an efficiency of 5%.

with respect to NaI(Tl), and a Nuclear Data 4096 channel analyzer. Figure 1 is a portion of the γ-ray spectrum obtained. By using a high-resolution detector, the 559-keV [76]As γ-ray was separated from the 564-keV γ-ray of [122]Sb.

The amount of arsenic or antimony in the samples was calculated from the following equation:

$$\text{p.p.m. As (or Sb)} = \frac{[\mu g \text{ As (or Sb) in std.}] \text{ (activity of sample)}}{\text{(weight of sample) (activity of std.)}}$$

The ratio of the activity of the sample to that of the standard was computed by the total peak area method[2].

Results and discussion

The results of the arsenic and antimony determinations by n.a.a. are shown in Table I. The results for arsenic obtained by n.a.a. are also compared with those obtained by X-ray fluorescence and by a colorimetric determination with arsine generation as reported by Schick and Watlington[3]. No data for antimony by other techniques were available.

TABLE I

ARSENIC AND ANTIMONY CONCENTRATIONS IN LAUNDRY AIDS DETERMINED BY n.a.a. AND COMPARATIVE ARSENIC RESULTS

Brand	Type[a]	p.p.m. Sb by n.a.a.[b]		p.p.m. As					
				N.a.a.[c]		X-ray[d]		Arsine[e]	
		Mean	s	Mean	s	Mean	s	Mean	s
A	EP	1	0.5	5	0.7	6	5	5	0.7
B	EP	2	0.5	17	2	11	3	15	2
C	EP	7	0.9	51	4	47	5	57	5
D	D	5	1	33	5	28	4	28	5
E	HDED	8	1	48	11	36	4	51	8
F	HDED	3	0.1	10	2	12	3	11	2
G	HDD	6	0.4	43	3	34	7	38	4

[a] EP. enzyme presoak; D. detergent; HDED, heavy-duty enzyme detergent; HDD. heavy-duty detergent.
[b] Mean and standard deviation from 3 determinations.
[c] Mean and standard deviation from 5 determinations.
[d] Mean and standard deviation from 3 determinations; data from Schick and Watlington[3].
[e] Mean and standard deviation from 3–6 determinations; data from Schick and Watlington[3].

Several problems were encountered in the early phases of this work. Originally, arsenic was determined radiochemically after neutron activation by a sulfide precipitation technique. However, because the detergent solution foamed excessively, results were unsatisfactory. The second and most critical problem was the non-resolution of the 559-keV γ-ray from arsenic and the 564-keV γ-ray from antimony. The radiochemical procedure originally used did not separate antimony from arsenic, and the NaI(Tl) detector originally used did not resolve the two γ-rays. The first determinations were unrealistically high for arsenic because of this

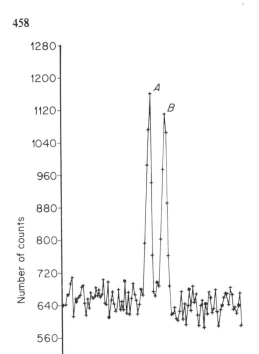

Fig. 1. A portion of the γ-ray spectrum of an irradiated laundry aid. (A) 559-keV ^{76}As; (B) 564-keV ^{122}Sb.

interference from antimony. By routinely checking the arsenic samples with a high-resolution Ge(Li) detector, the interfering γ-ray of antimony was resolved. All subsequent determinations were done instrumentally with the Ge(Li) detector, and both arsenic and antimony were determined.

The precision of the analyses may be estimated from Table I. Visual inspection of the samples and the variation between replicate analyses led us to believe that the samples may not have been homogeneous, which makes intercomparison of the different techniques difficult. However, such a comparison showed that the results for arsenic by n.a.a. agreed with those by the arsine and X-ray techniques within about 15%.

Other neutron-induced nuclear reactions leading to ^{76}As besides ^{75}As(n,γ)^{76}As have small cross-sections and do not interfere with the arsenic determinations[4,5]. The ^{122}Te(n,p)^{122}Sb reaction leads to ^{122}Sb but unless gross amounts of tellurium are present this interference is negligible[4]. Neither ^{122}Sb or ^{76}As is a fission product so that trace amounts of fissionable material would not interfere with these measurements. For the small sample size used and the phosphate matrix, self-shielding is negligible.

Neutron activation has certain advantages as a means of analysis for arsenic and antimony in laundry aids. Since the sample is sealed in a quartz vial at the beginning of the irradiation and the seal is not broken during the analysis, there is no chance of arsenic losses by volatilization. Because no reagents are used, blank errors from reagent contamination cannot occur. The technique used does not exploit special properties of the laundry aids and so would be applicable to a variety of matrices. Arsenic and antimony were not detected in the quartz vials. The analysis was done instrumentally so that the time expended per sample was brief (less than 15 min for the analysis of both elements).

The authors thank Mr. R. E. Simpson and Dr. R. M. Hehir of the Food and Drug Administration and Dr. Keith Marlow and the Reactor Staff of the Naval Research Laboratory for their help in various aspects of this work.

REFERENCES

1 E. E. Angino, L. M. Magnuson, T. C. Waugh, O. K. Galle and J. Bredfeldt *Science*, 168 (1970) 389.
2 P. A. Baedecker, *Anal. Chem.*, 43 (1971) 405.
3 A. L. Schick and P. Watlington, *FDA By-Lines*, 1 (1970) 79.
4 R. C. Koch, *Activation Analysis Handbook*, Academic Press, New York, 1960.
5 F. Baumgartner, *Table of Neutron Activation Constants*, Verlag Karl Thiemig Kg, München, 1967.

Look firstly at the title: ... instrumental NAA. We can infer from this that the technique did not involve wet chemical procedures, and hence that it was non-destructive.

Secondly, consider the concentration levels concerned. They are specified as 1–8 ppm Sb and 5–51 ppm As, ie the range 1–50 × 10^{-6} g Sb/As per gram of sample. When your read a little further and see that the sample size was about 500 mg you can see that the absolute levels of Sb/As being detected are in the range 0.5–25 × 10^{-6} g. You have seen previously that, for many elements, limits of detection much lower than this are attainable.

Now let us consider the experimental procedure adopted here. The samples are specified as being 'without boron', This is because boron has a very high capture cross-section for neutrons—roughly 150 times greater than for either arsenic or antimony.

If you happen to work in the detergent industry you will know that this is a severe restriction: many detergents do contain quite high concentrations of boron in the form of peroxyborates.

Next we note the quite small sample size, and the use of oxide and metal samples

∏   Why are oxides/metals specified as standards?

        You should remember from paragraph 2.1.1 that this is to minimise the chance of other stable isotopes giving radioactive products.

The description of the procedure emphasises the need for very careful experimental work to avoid contamination—this is a very sensitive technique. Also note that the samples were sealed.

Why? Because both As and Sb can give volatile compounds.

The irradiation conditions are interesting. The neutron flux was 10^{17} neutrons s^{-1} m^{-2}, which fits very well with the range we specified (4.4.3), and the irradiation time was quite short relative to $t_{0.5}$ for ^{122}Sb and ^{76}As.

Now we come to an important point: there was a pre-counting delay time of 3–5 days, 'to allow ^{24}Na activity to decrease relative to ^{122}Sb and ^{76}As'. Why might this be necessary? Firstly, of course, there will be quite high levels of sodium in any detergent (certainly far above the ppm level). Hence if the reaction

$$^{23}_{11}Na + ^{1}_{0}n \rightarrow ^{24}_{11}Na$$

has a similar value of σ to those for ^{75}As and ^{121}Sb, the induced activity would swamp any ^{122}Sb and/or ^{76}As activity.

SAQ 5.1a   The relevant data for the determination given in the paper are:

σ/m^2 for ^{23}Na $\rightarrow$ ^{24}Na, ^{75}As $\rightarrow$ ^{76}As, and ^{121}Sb $\rightarrow$ ^{122}Sb

$=$ 0.53×10^{-28}, 4.30×10^{-28}, and 3.90×10^{-28} respectively.

A for Na, As, and Sb are 22.99, 74.92, and 121.75 g respectively.

f for ^{23}Na, ^{75}As and ^{121}Sb are 1.0, 1.0, and 0.573 respectively.

Using the values of ϕ, $t_{0.5}$, and irradiation time given in the paper and taking $T = 4$ days, calculate the relative activities for a sample of detergent containing 10^4 ppm Na, 25 ppm As, and 5 ppm Sb.

The counting system fits in all respects with material we have covered earlier: a semiconductor detector with an excellent resolution was used. As you read further in the paper you should see why such a detector was necessary.

∏   What is the major advantage of semiconductor detectors? If you can't remember, look back to Paragraph 3.3.2. where you will see that we discussed their excellent resolution compared to that for scintillation detectors.

In this case the γ ray from ^{76}As is very close in energy to that from ^{122}Sb (559 keV and 564 keV respectively), and only a Ge(Li) detector has the capability to resolve such close γ rays. Fig. 1 of the paper shows how well this was achieved. By contrast, a scintillation detector would have given a single broad band covering both γ rays.

Now let us look at the calculation. Since we are comparing the activity induced in the unknown with that in the standard, there is no need to use the complicated equation for calculating absolute levels of induced activity, and so the calculation is much easier. We should also be careful about units.

∏   The arsenic content is quoted in ppm, and the weight of arsenic in the standard is specified in μg. In what units should we express the weight of sample?

      Since the usual definition is 1 ppm $= 10^{-6}$ g g^{-1} we must give the weight of sample in *grams*.

SAQ 5.1b   A standard containing 3 μg of arsenic gave 1200 counts min^{-1}, and a sample of the unknown weighing 0.4955 g gave 2081 counts min^{-1}. Calculate the arsenic content in the unknown.

                                                              $\longrightarrow$

SAQ 5.1b

The authors make an interesting point about possible interferences. Earlier in the learning text (Section 4.4.2) we discussed possible interference through overlapping γ rays from the various products of irradiation. In this case the authors considered the possibility of forming the identical isotope to the one being counted by reactions other than the one on which the determination is based. For example:

$$^{121}_{51}\text{Sb} + {}^{1}_{0}\text{n} \rightarrow {}^{122}_{51}\text{Sb} + \gamma \quad \text{Desired reaction.}$$

$$^{122}_{52}\text{Te} + {}^{1}_{0}\text{n} \rightarrow {}^{122}_{51}\text{Sb} + {}^{1}_{1}\text{p} \quad \text{Interfering reaction.}$$

Fortunately this second reaction has a negligible effect, but it is always necessary to check for this sort of interference.

The overall advantage of the method is summarised towards the end of the paper, namely the fact that it is non-destructive and hence free from any contamination.

Finally we should ask whether there are negative aspects to this very neat determination. There really is only one—the time needed. Although the authors claim that the time expended per sample was less than 15 minutes, this of course is for post-irradiation counting. Since there was a 3–5 day delay after counting this is the actual analysis time per sample. For routine work this might be considered too long a 'turn-round' time.

You might also be disturbed by the rather poor agreement between the NAA results and those from X-ray fluorescence and spectrophotometry, though this is not necessarily a fault of NAA.

5.1.2. Radioimmunoassay

The paper with which you are provided (P A Mason, B Law, K Pocock and A C Moffat, *Analyst* (1982), **107**, 629–33) illustrates very neatly the principles we discussed in Section 4.3.4, and is particularly useful because it deals with materials (ie barbiturates) with which, in general terms at least, many chemists are familiar, rather than with the much more complex systems for which RIA is often used.

At this point it would be a good idea for you to read the paper and get a general idea of the methods.

We are not concerned with the detail of this particular determination, but simply with the experimental aspects which illustrate the theory behind the method.

Let us consider first the radioactive tracer which is used in the procedure; it is a barbiturate derivative that is labelled with ^{125}I using the method specified in Section 4.3.4. ^{125}I emits low energy γ rays, and hence is easily counted; the paper specifies a γ counter. Next let us consider the specific activity of the labelled barbiturate. It is given in two sets of units: TBq mmol^{-1} and MBq μg^{-1}.

∏   What is the relationship between TBq and MBq?

See Section 1.4.1. 1 TBq, ie 1 terabecquerel is defined as 10^{12} disintegrations s^{-1} and 1 MBq, ie 1 megabecquerel as 10^6 disintegrations s^{-1}, so that 1 TBq = 10^6 MBq.

Analyst, June, 1982, Vol. 107, pp. 629–633 629

Direct Radioimmunoassay for the Detection of Barbiturates in Blood and Urine

P. A. Mason, B. Law, K. Pocock and A. C. Moffat

Home Office Central Research Establishment, Aldermaston, Reading, Berkshire, RG7 4PN

A radioimmunoassay has been developed for the detection of barbiturates in blood and urine without any pre-treatment of the sample. It is based on a radioiodinated derivative of 4-hydroxyphenobarbitone which allows use of relatively simple gamma-counting procedures.

The assay can detect therapeutic levels of barbiturates in very small amounts (50 μl) of blood and urine samples. It is cheap, rapid, simple to perform and is broadly specific for the barbiturate class of drugs to the exclusion of related drugs. The assay is, therefore, very well suited to the task of screening large numbers of samples for the presence of barbiturates.

Keywords: Barbiturate detection; blood; urine; radioimmunoassay

A radioimmunoassay for barbiturates was first described by Flynn and Spector in 1972.[1] Their assay was subsequently modified[2] and marketed in kit form by Hoffman-La Roche Inc. under the trade-name Barbiturate Abuscreen. This assay, which is based on a radioiodinated tracer (no details of the preparation of which were given), is reliable, rapid and easy to perform and can be used for the direct analysis of biological samples. The kit is still the only radioimmunoassay for barbiturates, other than specific assays for phenobarbitone, which has been reported in the literature. It has been evaluated by several groups of workers and its performance compared with other techniques such as thin-layer chromatography (TLC)[3] and gas - liquid chromatography (GLC) and TLC.[4–6]

An enzyme immunoassay for barbiturates in urine samples is marketed by Syva (UK) Ltd. under the trade-name Emit Barbiturate DAU. This assay has also been evaluated and its performance compared with that of other techniques, including the Abuscreen radioimmunoassay,[7] spectrophotometric assay[8] and TLC and fluorimetric assays.[9]

Both types of immunoassay compare well with other methods of analysis. Unlike the Abuscreen, use of the Emit assay is restricted to urine samples, which is a major drawback. The Abuscreen assay has now gained wide acceptance for barbiturate analysis. It does, however, have some disadvantages in its use as a screening test in toxicology laboratories. An attempt was therefore made to develop a similar radioimmunoassay.

Experimental

Materials and Equipment

The buffer used throughout the assay was 0.1 M phosphate buffer (pH 7.4) containing 0.2% m/V bovine gamma-globulin (Cohn Fraction II), 0.5% m/V bovine serum albumin (both from Sigma Chemical Co., Poole, Dorset) and 0.01% m/V sodium azide (BDH Chemicals, Poole, Dorset). Antiserum, obtained from an Emit Barbiturate DAU kit [Syva (UK) Ltd., Maidenhead, Berkshire] was diluted with assay buffer (1 + 29) as required.

The radiolabelled barbiturate derivative, 5-ethyl-5-(3-[^{125}I]-4-hydroxyphenyl)barbituric acid ([^{125}I]-4OHPB), specific activity 1.56 TBq mmol^{-1} (4.18 MBq μg^{-1}), was prepared by iodination of 5-ethyl-5-(4-hydroxyphenyl)barbituric acid by the chloramine-T reaction[10] as previously described[11] and stored in ethanol at 4 °C. It was diluted with assay buffer (1 + 149) as required, to give approximately 545 Bq (0.13 ng) per 100 μl. Solutions of quinalbarbitone were prepared in synthetic urine[12] at concentrations of 0, 25, 50, 100 and 200 ng ml^{-1}.

Method

Radioimmunoassay

Solutions of sample or standard (50 μl) were pipetted into duplicate sets of plastic, disposable microcentrifuge tubes. To these were added [^{125}I]-4OHPB and antiserum (100 μl of each).

The tubes were then capped, shaken and incubated at room temperature for 10 min. Equilibrium was shown to be stable for up to 24 h after that time. Saturated ammonium sulphate solution (250 μl) was then added and the tubes were re-capped, shaken and incubated at room temperature for at least 10 min before being centrifuged (1.5 min, 14 000 g). The supernatant was removed and the tubes containing the precipitates were counted for 1 min each in a gamma-counter. Urine and whole blood samples were diluted by a factor of 10 in synthetic urine and assayed as described above.

In order to determine the background levels of cross-reactivity, 50 samples each of blood and urine were obtained from normal subjects who were not receiving barbiturate-type medication and assayed as described above. The condition of the blood samples ranged from fresh unhaemolysed to haemolysed/putrefied.

A study was also carried out to determine whether the assay was affected by the presence of preservative agents. To known blank urine samples (2.5 ml) were added phenylmercury(II) nitrate and sodium fluoride (50 + 100 mg); sodium sulphate and sodium fluoride (300 + 300 mg); and sodium azide (50 mg). Blood samples (1 ml) were dispensed into vials containing sodium fluoride and potassium oxalate (37.5 + 18.7 mg) and sodium nitrite (25 mg). The concentration of preservatives in urine samples was approximately five times greater than that recommended for forensic use. The concentration of preservatives in the blood samples was approximately 2.5 times greater than that recommended.

Cross-reactivities of a number of barbiturates and structurally related compounds were determined (Table I).

Oral ingestion of barbiturates

Experiment 1. Five volunteers each took an oral dose of one barbiturate (Table II) in the morning (approximately 0900 h). Venous blood was obtained before and at 2 and 8 h after ingestion of the drug. The blood was stored in plastic vials containing EDTA (Sterilin Ltd., Teddington, Middlesex) at 4 °C. It was analysed by both radioimmunoassay and a high-performance liquid chromatographic (HPLC) procedure.[13]

Experiment 2. Five volunteers each took a single therapeutic dose of one barbiturate (Table III). Urine samples, which were obtained just before and up to 48 h after drug ingestion, were stored frozen prior to analysis. Both experiments were carried out on volunteers from our laboratory staff. The dosage level was approved by the Medical Ethics Committee, Chemical Defence Establishment, Porton Down.

Results and Discussion

The relative cross-reactivities of a number of barbiturates and structurally related compounds are shown in Table I. From these it is apparent that the antiserum has the necessary

TABLE I

CROSS-REACTIVITIES OF SEVERAL BARBITURATES, BARBITURATE METABOLITES
AND STRUCTURALLY RELATED COMPOUNDS

Compound	Relative reactivity* in radio-immunoassay	Compound	Relative reactivity* in radio-immunoassay
Cyclobarbitone	0.14	3-Hydroxybutobarbitone	1.1
Amylobarbitone	0.12	3-Hydroxyamylobarbitone	1.2
Butobarbitone	0.16	4-Hydroxyphenobarbitone	1.8
Quinalbarbitone	0.10	Glutethimide	>100
Pentobarbitone	0.16	Aminoglutethimide	>100
Phenobarbitone	0.26	Phenytoin	>100
Butalbital	0.17	Primidone	>100
Barbitone	0.45	Ethosuximide	>100
Methohexitone	0.49	Thiobarbituric acid	>100
N-Methylphenobarbitone	3.8	Caffeine	>100
3-Hydroxypentobarbitone	0.5	Theophylline	>100

* Amount in μg ml^{-1} that has the same degree of binding as 100 ng ml^{-1} of quinalbarbitone.

characteristics for development of an assay for screening samples for the presence of barbiturates: a good level of binding of the major members of the series and a very low level of binding of potentially interfering compounds such as primidone or phenytoin. However, primidone is metabolised to phenobarbitone, which will be detected by the assay. The antiserum also shows a high level of binding with hydroxylated metabolites as well as their parent barbiturates. This is a particularly important feature of an assay that is designed for use in forensic toxicology where urine samples are very common.

An example of the calibration graph obtained with the assay is shown in Fig. 1. The useful range of the graph is twice that of the Barbiturate Abuscreen so that fewer dilutions of each sample need to be assayed to ensure a successful result. The five commonly prescribed parent barbiturates have very similar cross-reactivities (Table I), so that a semi-quantitative result can be obtained without preparing a calibration graph for each drug.

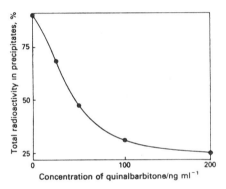

Fig. 1. Calibration graph for quinalbarbitone obtained using the barbiturate radioimmunoassay.

The distribution of background levels of cross-reactivity in 50 samples of urine from normal subjects who were not receiving barbiturate medication was positively skewed with a mean and standard deviation of 15 ± 28 ng ml^{-1}. The positive/negative cut-off for urine samples was set at 100 ng ml^{-1}, thus ensuring a $>99\%$ probability of obtaining a true positive result. A similar analysis of 50 blood samples gave a mean level of background cross-reactivity of 33 ± 39 ng ml^{-1}. The positive/negative cut-off was therefore set at 150 ng ml^{-1} for blood samples. Coefficients of variation for 50 and 100 ng ml^{-1} of quinalbarbitone in urine were 5.2% and 5.3% intra-assay ($n = 15$) and 9.7% and 11.8% inter-assay ($n = 12$).

No significant decrease in binding of the radiolabelled derivative to the antibody was caused by any of the preservatives studied, even at concentrations five times greater than those recommended for sample preservation.

As an alternative to the antiserum from the Emit Barbiturate DAU kit, an anti-phenobarbitone serum (Miles Laboratories, Slough, Berkshire) was also tested. Although this had very similar properties to the Emit antiserum, two major disadvantages were apparent; decreased binding of the hydroxylated barbiturate metabolites and unacceptably high coefficients of variation when used in an assay. Accordingly, the Emit antiserum was preferred.

In experiment 1 the assay was able to detect the levels of five common barbiturates in blood, resulting from a single therapeutic dose, for several hours after ingestion (Table II). The blood samples were also analysed by an HPLC procedure[13] for comparison with the radioimmunoassay and these results are also shown in Table II. The barbiturate levels are generally higher than those obtained by the HPLC method because the radioimmunoassay procedure also detects barbiturate metabolites whereas the HPLC method only detects the parent drugs. Overall, the agreement between the two methods was fairly good ($r = 0.806$).

TABLE II

BARBITURATE BLOOD LEVELS OF FIVE VOLUNTEERS FOLLOWING ORAL DOSES OF THE DRUGS

RIA = radioimmunoassay.

Approx. time after oral dose/h	Method	Barbiturate blood level*/μg ml^{-1}				
		Subject 1: cyclobarbitone (185 mg†)	Subject 2: butobarbitone (200 mg†)	Subject 3: amylobarbitone (182 mg†)	Subject 4: pentobarbitone (182 mg†)	Subject 5: quinalbarbitone (183 mg†)
2	RIA	1.10	2.82	2.58	5.62	5.08
	HPLC	0.86	3.32	2.29	3.15	1.99
8	RIA	2.83	2.46	1.91	3.41	4.37
	HPLC	2.95	2.66	1.75	2.26	1.57

* RIA calibration graph prepared with quinalbarbitone.
† Dose of free barbiturate.

The results from experiment 2 show that by using this assay the barbiturate levels in urine could be measured for at least 2 d following a single small therapeutic dose (Table III).

Obviously, therefore, the assay results must be treated with caution, as a positive result can be obtained long after the pharmacological effects have passed. In forensic practice, it is likely that samples will have to be diluted by a factor of 1000 in order to obtain a value that lies on the calibration graph.

If the antiserum is used at the dilutions specified, the cost per test is about 4.5 p, comparing very favourably with the Barbiturate Abuscreen, which costs approximately £1 per test. By using the antiserum at a greater concentration, the range of the calibration graph may be extended, but obviously this will increase the cost per test and cause a reduction in sensitivity

The assay that has been described fulfils the requirements for a screening test for barbiturates in biological samples. It is specific for barbiturates, requires a small volume of sample, is cheap and easy to perform and should prove extremely useful in both forensic and clinical toxicology.

TABLE III

BARBITURATE LEVELS IN URINE OF FIVE VOLUNTEERS FOLLOWING ORAL DOSE OF THE DRUGS

Approx. time after oral dose/h	Barbiturate urine level*/μg ml^{-1}				
	Subject 1: cyclobarbitone (185 mg)	Subject 2: butobarbitone (100 mg)	Subject 3: amylobarbitone (55 mg†)	Subject 4: pentobarbitone (45 mg†)	Subject 5: quinalbarbitone (55 mg)
2	0.56	0.42	0.31	0.50	0.75
4	10.97	0.59	NS‡	0.80	1.20
6	NS‡	0.74	0.54	1.15	1.60
8	11.83	0.90	0.34	1.30	0.97
12	13.77	0.43	0.54	0.75	0.77
24	4.74	0.94	0.45	2.20	1.60
48	2.29	0.95	0.37	1.08	0.90

* Calibration graph prepared with quinalbarbitone.
† Dose of free barbiturate.
‡ NS = sample not available.

We thank Dr. R. Gleadle (Chemical Defence Establishment, Porton Down) for his help in the collection of blood samples and Dr. R. Gill (Home Office Central Research Establishment) for the analysis of the blood samples by HPLC.

June, 1982   DETECTION OF BARBITURATES IN BLOOD AND URINE   **633**

References

1. Flynn, E. J., and Spector, S., *J. Pharmacol. Exp. Ther.*, 1972, **181**, 547.
2. Cleeland, R., Davis, R., Heveran, J., and Granberg, E., *J. Forensic Sci.*, 1975, **20**, 45.
3. Mulé, S. J., Whitlock, E., and Jukofsky, D., *Clin. Chem.*, 1975, **21**, 81.
4. Roerig, D. L., Lewand, D. L., Mueller, M. A., and Wang, R. I. H., *Clin. Chem.*, 1975, **21**, 672.
5. Jain, N. C., Budd, R. D., Sneath, T. C., Chinn, D. M., and Leung, W. J., *Clin. Toxicol.*, 1976, **9**, 221.
6. De Ferrari, F., Gambaro, V., Lodi, F., Marozzi, E., and Saligari, E., *Minerva Med.*, 1967, **67**, 2301.
7. Law, B., and Moffat, A. C., *J. Forensic Sci. Soc.*, 1981, **21**, 55.
8. Walberg, C. B., *Clin. Chem.*, 1974, **20**, 305.
9. Mulé, S. J., Bastos, M. L., and Jukofsky, D., *Clin. Chem.*, 1974, **20**, 243.
10. Hunter, W. M., and Greenwood, F. C., *Nature (London)*, 1962, **194**, 495.
11. Mason, P. A., and Law, B., *J. Labelled Compd.*, 1982, **19**, 357.
12. Rodgers, R., Crowl, C. P., Eimstad, W. M., Hu, M. W., Kam, J. K., Ronald, R. C., Rowley, G. L., and Ullman, E. F., *Clin. Chem.*, 1978, **24**, 95.
13. Gill, R., Lopes, A. T. T., and Moffat, A. C., *J. Chromatogr.*, 1981, **226**, 117.

Received *August 13th*, 1981
Accepted *November 23rd*, 1981

SAQ 5.1c   The paper states that the specific activity of the iodinated barbituric acid is 1.56 TBq mmol^{-1} or 4.18 MBq μg^{-1}. What is the molar mass of the compound expressed in grams?

We are not particularly concerned with the exact formulae of the compounds, nor of the antibody used (it is called the antiserum in this paper), but the concentrations and volumes are significant. The volumes specified are in the 50–100 μl range (ie 0.05–0.1 cm^3) with concentrations (see Fig. 1 of the paper) in the 0–200 ng cm^{-3} range. This compares very favourably with other sensitive analytical techniques for materials of this kind.

Finally we should notice the speed of the method, and the separation procedure. Equilibration was complete after 10 minutes, and the separation method (precipitation by ammonium sulphate followed by centrifugation) was also both simple and rapid. The authors certainly seem justified in the conclusion presented just above Table 3 of the paper.

5.1.3.   Substoichiometric Isotope Dilution Analysis

The paper with which you have been provided (R A Pacer and S M Benecke, *Analytical Chemistry* (1981), **53**, 1160–3) illustrates very well the principles we discussed in Section 4.3.2. When you read the introductory paragraphs you can see how well the authors describe the background to the technique, and particularly the need to avoid weighing minute amounts of radioactive derivatives. The experimental procedures raise several points of interest.

Firstly the radioactive isotope used (^{99}Tc) is of a different element to that being determined (Re). However, drawing on the chemical similarities of the elements, the authors are able to use derivatives of almost identical solubilities. The separation procedure (solvent extraction) is very easy indeed.

Next we can consider the properties of the radioactive isotope; the long half-life imposes no manipulative problems; there is no need to hurry the determination or to make corrections for decay.

What about possible counting procedures? The isotope is specified as emitting β^- particles of energy 0.292 MeV.

1160   *Anal. Chem.* **1981**, *53*, 1160–1163

Determination of Rhenium by Substoichiometric Pseudoisotopic Dilution Analysis with Technetium-99 and Liquid Scintillation Counting

Richard A. Pacer* and **Steven M. Benecke[1]**

Department of Chemistry, Indiana University–Purdue University, Fort Wayne Campus, Fort Wayne, Indiana 46805

Rhenium(VII) has been determined over a concentration range of 5×10^{-5} to 1×10^{-2} M by an isotope dilution analysis procedure, using $^{99}TcO_4^-$ as a radioisotopic tag. A substoichiometric amount of $(C_6H_5)_4AsCl$ is used as precipitant, followed by extraction into $CHCl_3$ and liquid scintillation counting. The average error over this concentration range was 3.3%. Solutions containing as little as 9 ppm Re can be readily determined. Replicate analyses yielded a coefficient of variation of 0.9%, while the average difference observed when vials were simply recounted was 0.7%. The following substances were found *not* to interfere, when present in 10-fold excess: $ZnCl_2$, $MnCl_2$, K_2CrO_4, $FeSO_4$, Na_2MoO_4, and NaF. Major interferants, causing low count rates and high apparent Re(VII) concentrations, include $KSCN$, KI, $NaClO_4$, KIO_4, $KMnO_4$, and $SnCl_2$.

A review of the past 10 years of *Chemical Abstracts* reveals no less than 24 different methods proposed for the determination of rhenium. The principal analytical methods by far are spectrophotometry and neutron activation analysis. Unfortunately, despite the simplicity of the technique, isotope dilution analysis per se is *not* included among the 24 methods (although, as Perezhogin (*1*) has shown, isotope dilution procedures may be incorporated into neutron activation methods for determining rhenium).

In this paper, a relatively simple procedure for determining rhenium by isotope dilution analysis will be described. With the advent of easily programmed microprocessor-controlled liquid scintillation counters, no operator attention is required during the collection of counting data. In fact, it is possible to program an instrument to count hundreds of samples overnight and then shut down automatically.

Radioisotope dilution analysis is predicated upon conservation of activity (DPM) when a known amount of a radioisotope, Wo, of known specific activity, So, is added to an unknown amount, Wx, of the same element (in nonradioisotopic form)

$$So\ Wo = S(Wo + Wx) \qquad (1)$$

where S is the specific activity of the mixture of the radioactive and nonradioactive forms of the element. However, in order to determine S, one needs to isolate a fraction of the mixture, $f \times (Wo + Wx)$, by precipitation, electrodeposition, solvent extraction, or other suitable means of phase formation. Isolation is necessary so that one can measure both activity (DPM) and amount (typically, mass) of the product to obtain specific activity. But if the amount of product isolated is extremely small, as in trace analysis, its mass cannot be measured accurately with the usual analytical balance (readable to ±0.0001 g). However, if one isolates the same number of moles of final product for all sample and standard solutions, specific activity is directly proportional to activity

[1] Present address: 1203 Enchanted Forest, South Bend, IN 46637.

Table I. Decay Properties of Several Commercially Available Isotopes of Rhenium

radioisotope	half-life	radiation energies, MeV
^{183}Re	70 days	EC (100%), γ 0.252, 0.081, 26 others
^{186}Re	88.9 h	EC (8%), β^- 1.072 (71%), 0.934 (21%), γ 0.137, 3 others
^{188}Re	16.7 h	β^- 2.12 (79%), 1.96 (20%), γ 0.1551, 13 others

(DPM), and it is only necessary to measure the radioactivity of the final product and not its mass. This can be accomplished by ensuring that the limiting reagent is *not* the analyte ion (ReO_4^- in this case) but the precipitating (complexing, etc.) reagent added (($C_6H_5)_4AsCl$ in this case). These principles of substoichiometric isotope dilution analysis were presented by Ruzicka and Stary (*2*) in 1961.

Several isotopes of rhenium are available commercially, including ^{183}Re, ^{186}Re, and ^{188}Re. Their decay properties are listed in Table I.

The relatively short half-lives of ^{186}Re and ^{188}Re precludes their convenient use in studies which may extend over weeks or months and requires corrections for decay. Frequent reordering or laboratory generation of the radionuclide would be necessary, in any case. ^{183}Re could not be measured as efficiently by liquid scintillation counting as could other β^--emitting nuclides, and, despite its longer half-life, only 3% of its activity would remain after storage for 1 year. On the other hand, ^{99}Tc ($t_{1/2} = 2.13 \times 10^5$ years, β^- 0.292 MeV) has been reported (*3*) to have been assayed by liquid scintillation counting with efficiencies greater than 94%.

The long half-life of ^{99}Tc precludes any need for decay corrections and permits indefinite storage in a laboratory without measurable loss of activity. By tagging a series of perrhenate-containing solutions with $^{99}TcO_4^-$ ion, we can carry out a technique known as "pseudoisotopic dilution analysis". Tölgyessy, Braun, and Kyrs (*4*) characterize *pseudoisotopic dilution analysis* by the fact that ... "the mutually diluting substances are not isotopic but only bear a sufficiently close chemical resemblance". They cite as examples of species determined/diluent pairs: $^{137}Cs/K$ (natural), insulin/^{131}I-iodinated insulin, and ^{250}Fm, $^{246}Cf/^{241}Am^{3+}$.

In the present study, the final product will consist of a mixture of the tetraphenylarsonium salts of the perrhenate and pertechnetate ions. To ensure that the same number of moles of final product is produced in all cases, we chose a procedure so that tetraphenylarsonium chloride is the limiting reagent.

$$ReO_4^-(aq) + (C_6H_5)_4As^+(aq) \rightarrow (C_6H_5)_4AsReO_4(s)$$

$$TcO_4^-(aq) + (C_6H_5)_4As^+(aq) \rightarrow (C_6H_5)_4AsTcO_4(s)$$

In a sense, then, both the TcO_4^- and ReO_4^- ions are competing for a limited supply of $(C_6H_5)_4As^+$ ions. The solubility products of $(C_6H_5)_4AsReO_4$ and $(C_6H_5)_4AsTcO_4$, 2.6×10^{-9} and

8.6 × 10⁻¹⁰ at ambient room temperature, respectively (5), are sufficiently close to each other for the two anions to compete effectively. Isolation of $(C_6H_5)_4AsReO_4$ and $(C_6H_5)_4AsTcO_4$ from the supernatant liquid is easily accomplished by virtue of the fact that these compounds are readily extractable into $CHCl_3$.

By counting an aliquot of the $CHCl_3$ layer, a well-defined relationship between ReO_4^- molarity and net CPM may be established. This relationship may then be used to estimate the ReO_4^- ion concentration in an unknown solution.

EXPERIMENTAL SECTION

Instrumentation. All counting was done with a Beckman LS 7000 liquid scintillation system, using Library Program No. 6. This is a two-channel program, actually designed for counting dual-labeled ³H–³²P samples. Nevertheless, it works well for ⁹⁹Tc and allows one to monitor spectral shifts by observing the relative count rate in each of the two channels.

Reagents. *Ammonium Pertechnetate.* An aqueous solution of NH_4TcO_4 was obtained from the Amersham Corp., Arlington Heights, IL. It was standardized gravimetrically, using tetraphenylarsonium chloride as precipitant.

Potassium Perrhenate, $KReO_4$, and Tetraphenylarsonium Chloride, $(C_6H_5)_4AsCl$. Both were obtained in reagent grade form from Alfa inorganics, Inc. Aqueous solutions were prepared with doubly distilled water.

Procedure. A 5.00-mL portion of $KReO_4$ solution is added to a separatory funnel, followed by addition of 1.00 mL of 7.71 × 10⁻⁵ M NH_4TcO_4 and thorough shaking. Next, 5.00 mL of 5.00 × 10⁻⁴ M $(C_6H_5)_4AsCl$ is added, after which funnels are again shaken thoroughly. For the extraction, 10.00 mL of reagent grade $CHCl_3$ is added. Separatory funnels are shaken for 2 min, and the layers are allowed to separate. The lower ($CHCl_3$) layer is drawn off into a small clean and dry Erlenmeyer flask and covered with a cork stopper. Counting is carried out by adding 10.00 mL of Beckman Ready-Solv HP scintillation cocktail and 250 μL of $CHCl_3$ solution to a vial and using the liquid scintillation counter as previously described. Samples were counted by using an automatic quench compensation counting mode and were counted using an edit code specifying a 50.0-min counting time or a 2.0% 2σ error in the channel having the fewer number of counts (whichever occurred first). In most cases, the net effect was to yield a total count rate for the two channels having an uncertainty due to counting statistics of 1% or less.

In order to establish the lowest detection limit for Re(VII) using this method, we revised the procedure by using lower and lower concentrations of $KReO_4$ and $(C_6H_5)_4AsCl$, until a clearly defined relationship between count rate and $KReO_4$ concentration was no longer evident.

In order to ascertain the degree of selectivity of the method, a 10-fold excess (in relation to the number of moles of $KReO_4$ present) of a potential interfering substance was added and its effect on the count rate, if any, was noted.

Sufficient replications were made so that the precision of the *method* could be evaluated and distinguished from the precision of the *counting procedure*.

RESULTS AND DISCUSSION

Precision. For estimation of the precision of the method, a solution 3.00 × 10⁻³ M in $KReO_4$ was run in replicate eight times. No potential interferant was present; i.e., the procedure as described in the first paragraph under "Procedure" was followed. The results may be summarized as

$$\bar{X} \pm s = 1657.8 \pm$$
14.6 counts/min (coefficient of variation = 0.88%)

In order to ascertain how reproducible the actual counting was itself, we ran a set of $KReO_4$ solutions, widely varying in concentration. Sample vials were counted in the order shown below, and then the entire set of eight vials was recounted. Differences in count rate for the same vial (in net counts per minute) were noted. The results are presented in Table II.

The data show that when vials are recounted, the average difference in count rate is well within 1%. The data also show

Table II. Reproducibility of Recounting Vials of Samples Tagged with NH_4TcO_4

$KReO_4$, mol/L	first count, net counts/min	second count, net counts/min	% difference
1.00 × 10⁻³	4187.4	4149.3	0.9
4.44 × 10⁻³	1176.3	1174.6	0.1
5.00 × 10⁻³	1057.8	1054.6	0.3
6.67 × 10⁻³	807.5	809.3	0.2
8.00 × 10⁻³	669.8	665.0	0.7
1.00 × 10⁻²	543.6	545.2	0.3
2.00 × 10⁻²	224.0	226.0	0.9
3.00 × 10⁻²	182.5	186.6	2.2
av			0.7

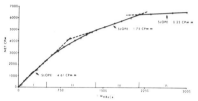

Figure 1. Net ⁹⁹Tc count rate (counts per minute) vs. the reciprocal of Re(VII) molarity, using 5.00 mL of 5.00 × 10⁻⁴ M $(C_6H_4)_4AsCl$ as precipitant.

that, as the molarity of $KReO_4$ increases, the count rate of the final product decreases. This is the expected effect, because the ReO_4^- and TcO_4^- ions are competing for a limited supply of $(C_6H_5)_4As^+$ ions. Thus fewer TcO_4^- ions (responsible for the observed activity) are incorporated into the $(C_6H_5)_4AsMO_4$ product (M = Re or Tc) as the $[ReO_4^-]/[TcO_4^-]$ ratio in the initial solution increases.

Accuracy. In order to evaluate whether the effect described above could be used to determine the molarity of the ReO_4^- ion in an "unknown" solution, an extensive series of $KReO_4$ solutions were run, ranging in concentration from 3.00 × 10⁻² to 3.33 × 10⁻⁴ M, using the procedure described previously. The upper concentration limit was selected simply because it comes close to the solubility limit of $KReO_4$ reported in the literature (6), 3.4 × 10⁻² M at 19 °C. A plot was made of net counts per minute as a function of the reciprocal of the $KReO_4$ molarity. The results are shown in Figure 1. The curve may be considered as consisting of four distinct regions, two of which may be approximated fairly well by straight line segments. These two regions, noted as regions I and III on the curve, are useful analytically. Region I is characterized by 15 points and has a slope of 4.61 counts/min M and an intercept of 76.2 counts/min. Region II is too extensively curved to be useful analytically. Region III, the second of the linear segments, is characterized by six points and has a slope of 1.75 counts/min M and an intercept of 2591 counts/min. Region IV is the "levelling-off" region. At concentrations in this region, the principle of substoichiometry no longer applies. That is, tetraphenylarsonium chloride is no longer the limiting reactant and therefore one can no longer assume that a constant number of moles of $(C_6H_4)_4AsMO_4$ is being transferred to the $CHCl_3$ layer. Thus, in order to determine ReO_4^- in solutions of Re(VII) molarity less than ~4.6 × 10⁻⁴ M, it is necessary to restore the principle of substoichiometry. This can be done, of course, by using a smaller amount of $(C_6H_5)_4AsCl$ in each run.

To evaluate the analytical error in regions I and III of the composite curve, a Hewlett-Packard calculator plotter was used to prepare a linear least-squares plot for each region and

1162 • ANALYTICAL CHEMISTRY, VOL. 53, NO. 8, JULY 1981

Table III. Evaluation of Percent Error in Region I of Composite Curve (Figure 1)

point no.	[KReO$_4$], M (actual)	net counts/ min	[KReO$_4$], M (calcd)	% error
1	3.00×10^{-2}	184.6	4.25×10^{-2}	41.7
2	2.00×10^{-2}	225.0	3.10×10^{-2}	55.0
3	1.00×10^{-2}	544.4	9.85×10^{-3}	- 1.5
4	8.00×10^{-3}	667.4	7.80×10^{-3}	-2.5
5	6.67×10^{-3}	808.4	6.30×10^{-3}	-5.5
6	5.00×10^{-3}	1042.0	4.77×10^{-3}	-4.6
7	4.44×10^{-3}	1175.4	4.19×10^{-3}	-5.6
8	4.09×10^{-3}	1254.5	3.91×10^{-3}	-4.4
9	3.85×10^{-3}	1326.1	3.69×10^{-3}	-4.2
10	3.26×10^{-3}	1472.0	3.30×10^{-3}	1.2
11	3.00×10^{-3}	1515.1	3.20×10^{-3}	6.7
12	2.65×10^{-3}	1816.3	2.65×10^{-3}	0.0
13	2.28×10^{-3}	2091.5	2.29×10^{-3}	0.4
14	2.10×10^{-3}	2258.7	2.11×10^{-3}	0.5
15	2.00×10^{-3}	2372.9	2.01×10^{-3}	0.5

Table IV. Evaluation of Percent Error in Region III of Composite Curve (Figure 1)

point no.	[KReO$_4$], M (actual)	net counts/ min	[KReO$_4$], M (calcd)	% error
1	7.27×10^{-4}	4983.7	7.31×10^{-4}	+0.6
2	6.56×10^{-4}	5278.3	6.51×10^{-4}	-0.8
3	5.88×10^{-4}	5594.9	5.83×10^{-4}	-0.9
4	5.48×10^{-4}	5798.3	5.46×10^{-4}	-0.4
5	5.00×10^{-4}	6030.2	5.09×10^{-4}	+1.8
6	4.56×10^{-4}	6395.0	4.60×10^{-4}	-1.1

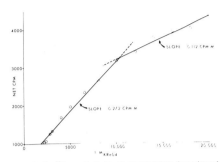

Figure 2. Net ^{99}Tc count rate (counts per minute) vs. the reciprocal of Re(VII) molarity, using 5.00 mL of 5.00×10^{-5} M (C$_6$H$_5$)$_4$AsCl as precipitant.

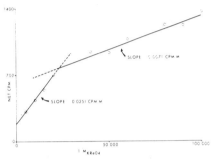

Figure 3. Net ^{99}Tc count rate (counts per minute) vs. the reciprocal of Re(VII) molarity, using 5.00 mL of 1.00×10^{-5} M (C$_6$H$_5$)$_4$AsCl as precipitant.

to calculate the respective slope and intercept of each. With these values, it is then possible to calculate the perrhenate ion molarity corresponding to a given observed net counts/min value. By comparing the *calculated* with the *actual* molarities, and noting the percent difference, an estimate of the analytical error for each region may be noted. The results for regions I and III are presented in Tables III and IV, respectively.

The first two points in Table III have unfavorable counting statistics, in view of the fact that a typical background count rate ranges from 45 to 55 counts/min. It may also be that the ReO$_4^-$/TcO$_4^-$ mole ratio is too large (1950 and 1300 for points 1 and 2, respectively) for the ReO$_4^-$/TcO$_4^-$ competition to be effective. The average error (average of absolute values) for the remaining 13 points is 2.9%.

The data in Table IV have an average error (average of absolute values) of 0.9%. In this region the ReO$_4^-$/TcO$_4^-$ mole ratio ranges from 47 (point 1) to 30 (point 6).

Sensitivity. It became obvious that, in order to ascertain the lowest concentration of ReO$_4^-$ which could be determined by this method, the region of substoichiometry would need to be extended by using smaller and smaller amounts of the limiting reagent, (C$_6$H$_5$)$_4$AsCl. Hence, in the next set of runs the molarity of (C$_6$H$_5$)$_4$AsCl was reduced by a factor of 10 (5.00 mL of 5.00×10^{-5} M (C$_6$H$_5$)$_4$AsCl was used in the procedure as described earlier). The results are shown in Figure 2. As with Figure 1, a smooth curve is obtained which may be approximated by two linear regions. The higher concentration region has a slope of 0.272 counts/min M, while the lower concentration region has a slope of 0.112 counts/min M. If an average percent error (average of absolute values) is calculated as described previously, it is seen to be 5.4% for the higher concentration linear region and 3.4% for the lower concentration linear region.

Continuing further along the same vein, 5.00 mL of 1.00×10^{-5} M (C$_6$H$_5$)$_4$AsCl was used in the next set of runs. Results are shown in Figure 3. The high concentration linear region

pertains to a KReO$_4$ molarity ranging from 2.00×10^{-4} to 5.00 $\times 10^{-5}$ M, with an average error of 3.6%. The lower concentration linear region pertains to a KReO$_4$ molarity ranging from 2.50×10^{-5} to 1.00×10^{-5} M and individual points have an average error of 12.2%. At still lower concentrations of (C$_6$H$_5$)$_4$AsCl, the relationship between net counts per minute and reciprocal of KReO$_4$ molarity becomes ill-defined. In view of the fact that a 12.2% error is excessively high, one may conclude that the lower concentration limit for which this method is applicable is 5×10^{-5} M ReO$_4^-$ (9.3 ppm Re).

Selectivity. There are a number of ways in which a chemical substance might interfere with this method. Species which form a precipitate with (C$_6$H$_5$)$_4$As$^+$ or with ReO$_4^-$ or TcO$_4^-$ ions would be expected to interfere (unless K_{sp} values were relatively large). Likewise, species which reduce TcO$_4^-$ and/or ReO$_4^-$ would be expected to interfere. With this in mind, a number of common chemical substances were evaluated to obtain a general picture of the selectivity (or lack thereof) of the method. The general analytical procedure described earlier was used, with the following modifications: 5.00 mL of 3.00×10^{-3} M KReO$_4$ was used in all runs and, prior to the addition of (C$_6$H$_5$)$_4$AsCl, 5.00 mL of a solution which was 3.0×10^{-2} M in the potential interferant was added (for the controls, 5.00 mL of redistilled water was used instead).

A pooled standard deviation was calculated for the controls and the replicate runs. The pooled standard deviation, based on 21 degrees of freedom, was found to be 14.9 counts/min. [The separate results for the eight control runs have already

ANALYTICAL CHEMISTRY, VOL. 53, NO. 8, JULY 1981 • **1163**

Table V. Summary of Data for Species Evaluated for Possible Interference in the Isotope Dilution Method for Rhenium[a]

group	substance	av net counts/min [b]
1	$ZnCl_2$	1664.5
	[dist H_2O (control)]	1657.8 (av of 8 values)
	$MnCl_2$	1656.7
	K_2CrO_4	1656.3
	$FeSO_4$	1652.4
	Na_2MoO_4	1650.4
	NaF	1649.9 (1640.5, 1663.4, 1645.7)
2	Na_2WO_4	1634.2 (1625.0, 1645.5, 1632.1)
	$CdCl_2$	1630.8 (1609.0, 1631.8, 1651.7)
	$NaC_2H_3O_2$	1629.3 (1636.3, 1626.4, 1625.2)
3	$CuCl_2$	1627.9 (1630.9, 1625.8, 1626.9)
	$BaCl_2$	1623.8 (1639.1, 1633.1, 1599.1)
4	KNO_3	1588.1 (1579.9, 1609.8, 1574.7)
	Na_3PO_4	1578.5
	$SnCl_4$	1568.6
	KBr	1565.4
	$Pb(NO_3)_2$	1546.5
	$Fe(NO_3)_3$	1542.0
5	KSCN	941.6
	KI	832.7
	$NaClO_4$	457.2
	KIO_4	363.1
	$KMnO_4$	75.5
	$SnCl_2$	0

[a] Listed in order of decreasing net counts/min. [b] Individual values in parentheses if more than one.

been presented ($\bar{X} = 1657.8$ counts/min, $s = 14.6$ counts/min); See Precision section].

For 21 degrees of freedom, values of t for the 95 and 99% confidence levels are 2.080 and 2.831, respectively. Thus, for substances run in triplicate and considered as possible interferants, rejection criteria (C) for 95 and/or 99% confidence may be readily calculated

$$C = \pm ts \sqrt{\frac{N_1 + N_2}{N_1 N_2}} \quad (2)$$

$$C_{95} = \pm(2.080)(14.9)\sqrt{\frac{8+3}{8 \times 3}} = \pm 21.0 \text{ counts/min} \quad (3)$$

$$C_{99} = \pm(2.831)(14.9)\sqrt{\frac{8+3}{8 \times 3}} = \pm 28.6 \text{ counts/min} \quad (4)$$

The results are summarized in Table V. The various substances tested can be classified into five groups, as follows:

Group 1. There is *no evidence that* any of these *substances interfere.* The counting rate is well within one standard deviation of the control mean value.

Group 2. These are *borderline cases.* Compared to the control mean value, these substances have counting rates which are significantly different at the 95% confidence level but *not* at the 99% confidence level.

Group 3. These substances all yield counting rates which are *slightly low* ($\sim2\%$ low), but sufficiently low to be significant at the 99% confidence level.

Group 4. These substances yield *moderately low* (4–7% low) counting rates and therefore interfere in the method.

Group 5. These substances constitute *major interferences* and yield counting rates which are 40% to 100% low in comparison to the control group.

Note. It should be kept in mind that, because of the inverse relationship between counting rate and ReO_4^- molarity, substances which cause *low counting rates* will yield *high results* for the apparent concentration of ReO_4^- ion in solution.

LITERATURE CITED

(1) Perezhogin, G. A. *Zavod. Lab.* **1965**, *31*, 402.
(2) Ruzicka, J.; Stary, J. *Talanta* **1961**, *8*, 228–234.
(3) Pacer, R. A. *Int. J. Appl. Radiat. Isot.* **1980**, *31*, 731–736.
(4) Tölgyessy, J.; Braun, T.; Kyrs, M. "Isotope Dilution Analysis"; Pergamon: Oxford, 1972; Chapter 1.
(5) Pacer, R. A. *Talanta* **1980**, *27*, 689–692.
(6) Wilke-Dorfurt, Von E.; Gunzert, T. *Z. Anorg. Allg. Chem.* **1933**, *215*, 369–387.

RECEIVED for review February 6, 1981. Accepted April 4, 1981. This work was supported in part by a grant from the Indiana University–Purdue University at Fort Wayne Research and Instructional Development Support Program. Acknowledgment is made to the donors of The Petroleum Research Fund, administered by the American Chemical Society, for partial support of this research. Presented in part before the Division of Analytical Chemistry at the 180th National Meeting of the American Chemical Society, Las Vegas, NV, Aug 1980. Presented in part before the Division of Analytical Chemistry at 181st National Meeting of the American Chemical Society, Atlanta, GA, Mar/Apr 1981 (accepted for presentation 12-24-80).

∏   Given that liquid samples are to be counted what are the possible counting methods?

From Sections 3.1 and 3.2 you will recall that the most likely methods are liquid Geiger counting and liquid scintillation counting, the choice normally depending on the β^- energy. For the isotope used in this paper the energy is quite low, and the efficiency of a liquid Geiger counter would certainly be less than the 94% quoted for liquid scintillation counting. Not surprisingly, the authors chose the latter.

So how does this procedure compare with what we said in Section 3.2.3? The isotope to be counted is extracted into chloroform so there are none of the problems associated with aqueous samples, and the organic sample is simply mixed with a scintillation 'cocktail' as we suggested (3.2.3). The counting rates (Tables II–IV and Figs. 1–3 of the paper) are entirely typical of tracer experiments.

Finally we can note the limit of detection for this method; this is quoted as 5×10^{-5} mol dm^{-3} of ReO$_4^-$ or 9.3 ppm Re.

SAQ 5.1d   Given that the relative atomic mass of rhenium is 186.21, calculate the concentration of Re in ppm of a solution of ReO$_4^-$ of concentration 5×10^{-5} mol dm^{-3}.

In summary this is a very neat and sensitive determination of rhenium with a comprehensive survey of the variables involved.

SUMMARY AND OBJECTIVES

Summary

Neutron activation analysis, radioimmunoassay, and substoichiometric isotope dilution analysis are considered to be the most important radioanalytical techniques. Each technique is illustrated by one published paper, and the papers are used to interpret and amplify theoretical principles and experimental procedures described earlier.

Objectives

You should now be able to:

- appreciate the significance of typical published examples of radioanalytical methods;

- relate the experimental procedures given in the papers to principles covered in earlier sections;

- perform calculations using the data in the papers.

5.2. POSSIBLE APPLICABILITY OF RADIOCHEMICAL METHODS

Overview

This brief section summarises the general principles which underlie the application of radioactive isotopes in analytical chemistry, and outlines the main areas of use. It also gives a checklist of questions to be satisfactorily answered before work begins.

5.2.1. Possible Applicability of Radiochemical Methods to New Analytical Situations

In order to summarise the ways in which radioanalytical methods can be helpful we need firstly to consider the range of problems that can be presented to an analyst. Broadly speaking we could say that the analyst is concerned with the methods of analysis, and with analyses themselves. Thus we could distinguish between the development of new methods and the validation of published methods on the one hand, and the carrying out of an analysis to measure one or more components of a material on the other.

You have seen that the areas of method development that are most amenable to the use of radioactive isotopes are separation and concentration procedures. The determination of the solubility of compounds and the investigation of possible decomposition reactions in analytical processes are others. It is almost impossible to generalise on such investigations; the limitation is much more likely to be availability of a suitable isotope rather than a particular property of the method being investigated.

The direct comparison of radioanalytical methods with 'conventional' methods is difficult also, because so much depends on the nature of the material being analysed, the number of components to be measured and the concentration levels at which measurements are required. It simply is not possible to say that a given radioanalytical method is 'better' than a non-radioactive one, without discussing all the specifications of the analysis. We shall therefore only make a few generalisations.

The simple application of isotope dilution procedures has many advantages over conventional methods, most often in terms of speed and limits of detection. The reason such procedures are not more widely used is attributable principally to the limited availability of the specialised laboratory facilities necessary for radioactive work.

Radioimmunoassay has achieved a pre-eminent position in many areas of analytical chemistry bordering on clinical chemistry; it can justly be said to have revolutionised many diagnostic procedures.

Activation analysis, typically with neutrons, has many advantages compared to conventional methods, particularly now that semiconductor detectors allow many analyses to be completely instrumental and non-destructive. Limits of detection are generally very good, though these need to be compared with those from conventional methods on an individual element basis. The other great advantage of activation analysis is the ability to analyse highly complex mixtures; it is quite common to read of the simultaneous determination of fifteen to twenty elements in a sample. The disadvantage is the need for irradiation facilities, normally with a high flux, and despite the development of ^{252}Cf sources this is likely to remain a problem.

It is convenient to conclude with a checklist of questions which you would need to answer if you were put in the position of needing to decide on the use of a radioactive isotope or the implementation of a radioanalytical method.

> Is a particular isotope needed, or simply a given type of radiation?

> If a particular isotope is needed, is it available?

> Is the isotope available in a suitable chemical form?

> If not, can it easily be converted to the right form?

> What are the physical characteristics of the isotope, particularly in relation to half-life and decay scheme?

Is there a convenient way of detecting and measuring the activity of the isotope?

What activity levels will be appropriate?

What safety precautions will be necessary in handling the isotope?

Is there a convenient disposal method after use?

If the activity is to be obtained by irradiation of an inactive sample is a suitable means of irradiation available?

Is the nature of the sample compatible with irradiation?

What levels of activity will be induced in the sample?

At what stage can it be safely handled after irradiation?

Whichever kind of application is being considered, is there an appropriate resource provision in terms of laboratory accommodation, equipment for counting, safety measurements, and trained staff?

Finally, though over-ridingly, do any apparent benefits of a radioactive method outweigh the hazards, however slight the latter may be?

SUMMARY AND OBJECTIVES

Summary

The relationships between analytical processes and their investigation using radiochemical methods are considered. The main criteria for choice of isotope and method are listed.

Objectives

You should now be able to:

- interpret the requirements of a given analytical situation in relation to its possible solution by radioactive means;

- itemise the experimental factors affecting a given method;

- come to a conclusion on the correct procedure to be followed.

Self Assessment Questions and Responses

SAQ 1.1a

It is instructive to try to relate the radius of the nucleus to that of the atom in more conventional terms. Imagine that the nucleus has a radius of 1 mm. What is the radius of the atom in units of metres?

Response

You should have obtained an answer of 10^2 m. If the radius of the atom is 10^5 times that of the nucleus it would on this scale be 10^5 mm ie 10^4 cm or 10^2 m. Thus if a pin-head on the table in front of you has a radius of 1 mm, and represents a nucleus, the most distant electrons of the atom are 100 m (say 100 yards) away! This should give you some idea of the size of a nucleus relative to that of an atom, and—incidentally—an appreciation that the bulk of an atom is space which is rather sparsely filled with electrons.

****

SAQ 1.1b

(*i*)   Calculate the volume of the nucleus, given that its radius is 10^{-15} m.

(*ii*) If this nucleus contains 20 neutrons and 20 protons each of mass 1 m_u (1 m_u = 1.66 × 10^{-27} kg) calculate the mass of the nucleus.

(*iii*) Hence determine the density of the nucleus.   ⟶

Response

(*i*) $\text{Vol} = \frac{4}{3}\pi r^3$

$= \frac{4}{3}\pi \times 10^{-45} = 4.189 \times 10^{-45} \text{ m}^3$

(*ii*) Total mass of nucleus $= 40 \times 1.66 \times 10^{-27} \text{ kg}$

$= 6.64 \times 10^{-26} \text{ kg}$

(*iii*) Density $= \text{Mass/Vol} = 6.64 \times 10^{-26}/4.189 \times 10^{-45}$

$= 1.585 \times 10^{19} \text{ kg m}^{-3}$

$= 1.585 \times 10^{16} \text{ g cm}^{-3}$

****

SAQ 1.1c	Predict, on the basis of $N:Z$ ratio, whether the following isotopes will be stable or radioactive, by ringing the appropriate responses:

$^{57}_{25}\text{Mn}$ S   R

$^{37}_{19}\text{K}$ S   R

$^{71}_{31}\text{Ga}$ S   R

$^{108}_{44}\text{Ru}$ S   R

$^{25}_{12}\text{Mg}$ S   R

$^{82}_{38}\text{Sr}$ S   R $\longrightarrow$

Response

$^{57}_{25}$Mn has 32n + 25p, so $N:Z$ = 1.28. The maximum value for stability is 1.25, so $^{57}_{25}$Mn is radioactive.

$^{37}_{19}$K has fewer neutrons than protons and so it must be radioactive.

$^{71}_{31}$Ga has 40n + 31p, so $N:Z$ = 1.29. Isotopes of elements in this range are radioactive only if $N:Z$ is greater than 1.40, so $^{71}_{31}$Ga is stable.

$^{108}_{44}$Ru has 64n + 44p, so $N:Z$ = 1.45. Again, the limiting value is 1.40, so $^{108}_{44}$Ru is radioactive.

$^{25}_{12}$Mg has 13n + 12p, so $N:Z$ = 1.08. Isotopes of elements in this range and are radioactive only if $N:Z$ is greater than 1.25, so $^{25}_{12}$Mg is stable.

$^{82}_{38}$Sr is the catch! The value of $N:Z$ is 1.16 which is less than 1.40, so you possibly predicted stability. In fact it is radioactive, because the ratio is much *less* than the minimum stable value for strontium.

****

SAQ 1.1d

Calculate E_B for $^{27}_{13}$Al given that the measured isotopic mass = 26.981 539 m_u, proton mass = 1.007 276 m_u neutron mass = 1.008 665 m_u, electron mass = 0.000 548 58 m_u, and confirm that E_B/number of nucleons lies in the expected range. $\longrightarrow$

Response

$^{27}_{13}Al$ has 13 protons, 13 electrons, and 14 neutrons. Hence the theoretical mass = 27.223 042 m_u. When the measured mass is subtracted the mass defect = 0.241 503 m_u. When this is multiplied by 931.5 MeV the binding energy = 224.96 MeV.

There are 27 nucleons, so that binding energy per nucleon = 224.96/27 = 8.33 MeV which is within the specified range 7.5–8.8 MeV.

****

SAQ 1.2a

The following statements refer to the α emitting isotope $^{239}_{94}Pu$ or to its α particles. Indicate whether the statements are true (T) or false (F).

(*i*) All α particles from $^{239}_{94}Pu$ have the same energy.

T / F

(*ii*) α spectrometry is a possible means of identifying $^{239}_{94}Pu$.

T / F

(*iii*) The neutron : proton ratio is lower in $^{235}_{92}U$ than in $^{239}_{94}Pu$.

T / F

(*iv*) α particles from $^{239}_{94}Pu$ travel long distances in body tissue.

T / F

(*v*) α particles from $^{239}_{94}Pu$ are *not* easily measured by Geiger counters.

T / F.

$\longrightarrow$

Response

(*i*) is true: α particles emitted from a particular decay process have a characteristic energy.

(*ii*) is true: α spectrometry is widely used for ^{239}Pu.

(*iii*) For $^{239}_{94}$Pu, n:p = 1.543 and for $^{235}_{92}$U, n:p = 1.554. Hence this statement is false.

(*iv*) You should remember that α particles travel very short distances in materials denser than air. Body tissue is in this category, so this statement is false.

(*v*) is true.

****

SAQ 1.2b

The following statements refer to the β^- emitting isotope $^{32}_{15}$P or to its negatrons.

(*i*) All negatrons from ^{32}P have the same energy.

(*ii*) Geiger counting is not a possible means of measuring ^{32}P.

(*iii*) The neutron:proton ratio is lower in $^{32}_{16}$S than in $^{32}_{15}$P.

(*iv*) Negatrons from ^{32}P are a greater external health hazard than most α particles.

(*v*) The product of β^- emission from $^{32}_{15}$P is $^{32}_{14}$Si.

Work out which *three* of these statements (*i*)–(*v*) are *false*. $\longrightarrow$

Response

(*i*)   is false.

(*ii*)   is false: Geiger counting *is* the standard method for ^{32}P.

(*iii*) For ^{32}S n : p = 1.00 and for ^{32}P n : p = 1.13. Hence this statement is true.

(*iv*)   The information in the text strongly suggests this is true.

(*v*)   When a neutron is converted to a proton the daughter isotope has a higher atomic number (by 1) than the parent. Hence this statement is false.

****

SAQ 1.2c	The decay scheme for $^{38}_{17}$Cl may be represented as:
	β^-/MeV : 1.11 (31%), 2.71 (16%), 4.81 (53%)
	γ/MeV : 1.60 (31%), 2.10 (47%)
	Draw this scheme diagrammatically, plotting energy on a vertical scale and atomic number on a horizontal scale, and hence identify the daughter product. $\longrightarrow$

Response

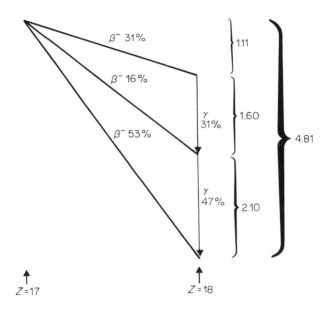

Fig. 1.2d. *Decay Scheme for $^{38}_{17}Cl$*

For β^- emission $^1_0n \rightarrow {}^1_1p + {}^0_{-1}\beta$

Therefore, Product is $^{38}_{18}Ar$

If you could not convert the data in the question to the above diagram, look again at Fig. 1.2c for ^{130}I bearing in mind the need not just for there to be a total of 100% emissions, but for the individual energy changes to add up to the overall energy decrease.

****

SAQ 1.3a	The activity of a sample of a radioactive isotope was found to be 5000 disintegrations per second when initially measured, and 90 seconds later was found to be 1500 disintegrations per second. What are the value and the units of the decay constant?

Response

The decay constant, λ, can be calculated from Eq. (1.3c):

$A = A_0 e^{-\lambda t}$.

$A_0 = 5000$ disintegration s^{-1}

$A = 1500$ disintegration s^{-1}

$t = 90$ s

$\therefore$ Putting these figures in Eq. 1.3c

$1500 = 5000\ e^{-\lambda 90}$

$\therefore$ $1500/5000 = e^{-\lambda 90}$

Before we go further we should note that for the ratio A/A_0 the units cancel.

$\therefore$ $0.3 = e^{-\lambda 90}$

Taking logarithms to base e:

$\ln 0.3 = -\lambda 90$

$-1.2 = -\lambda 90$

$\therefore$ $\lambda = 1.2/90$

$\therefore$ $\lambda = 0.0134$

What are the units of λ? t was expressed in s, and since t is in the denominator of the above equation λ must be in units of s^{-1}.

****

SAQ 1.3b

The data below refer to the decay of a sample of ^{234}Pa.

A/disintegrations s^{-1}	237	184	160	129	111	91	77	63
t/s	0	26	42	62	80	100	116	138

By drawing a suitable graph of these data calculate:

(i) the decay constant and

(ii) the half-life for this isotope. $\longrightarrow$

Response

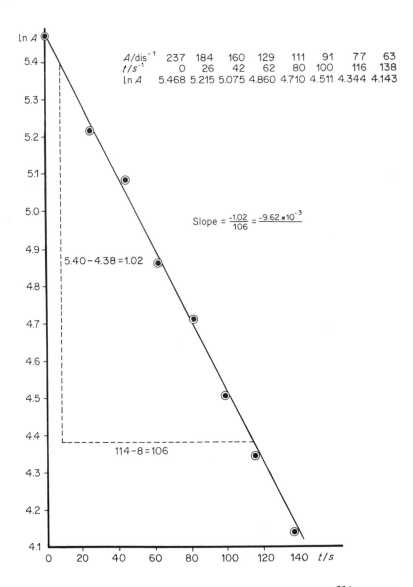

Fig. 1.3c.   *Graph of ln A plotted against t for* ^{234}Pa.

By plotting ln A against t you should get a straight line of slope -9.62×10^{-3} s^{-1} and hence $\lambda = 9.62 \times 10^{-3}$ s^{-1} (see Fig. 1.3c). If you plotted log A against t, the slope should have been -4.18

$\times$ 10^{-3} s^{-1} and since the slope $= -\lambda/2.303$ you should again have obtained $\lambda = 9.62 \times 10^{-3}$ s^{-1}.

Since $t_{0.5} = 0.693/\lambda$ your value of $t_{0.5}$ should be 72.0 s.

If your value is different from this check the values you have obtained for log A (or ln A), slope, and λ.

****

SAQ 1.3c

The half-life of the isotope $^{24}_{11}$Na is 15.0 h. The disintegration rate of a sample of this isotope was 16 000 disintegrations min^{-1} at 09.00 h on a certain day. At 21.00 h on the following day the disintegration rate (disintegrations min^{-1}) was:

(*i*) 9191;   (*ii*) 13 548;   (*iii*) 3032.

Response

The correct answer is (*iii*):

$t_{0.5} = 15.0$ h and hence $\lambda = 0.693/15 = 0.0462$ h^{-1}.

ln $A =$ ln $A_0 - \lambda t$

$\quad\quad =$ ln $16000 - 0.0462 \times 36$

(since 09.00 on day 1 to 21.00 on day 2 $= 36$ h)

$\therefore$ ln $A = 9.6803 - 1.6632$

$\quad\quad\quad = 8.0171$

$\therefore$ $A \quad = 3032$ disintegrations min^{-1}

If you obtained (*i*) as the correct answer you must have taken *t* = 12 h (ie 21.00 h on the *same* day as the initial measurement).

If you obtained (*ii*) you made a decimal point error in calculating λ, and used λ = 0.00462 h^{-1}.

If you obtained none of these three values, check through the above calculation, and repeat it for a reading taken at 06.00 h on the second day. You should then obtain *A* = 2000 disintegrations min^{-1}.

****

SAQ 1.4a
> A sample gave an experimental count rate of 9800 counts minute^{-1} in a counter of 30% efficiency. Choose from options (*i*)–(*iv*) the activity expressed as Bq and Ci respectively:
>
> (*i*)   1.96 MBq or 52.97 µCi
>
> (*ii*) 544.44 Bq or 14.71 nCi
>
> (*iii*) 32.67 kBq or 0.88 µCi
>
> (*iv*) 9.80 kBq or 0.26 µCi

Response

The actual count rate 9800 counts minute^{-1} in a counter of 30% efficiency is 9800 × 100/30 disintegrations minute^{-1}.

= 32666.67 disintegrations min^{-1}

or 32666.67/60 disintegrations s^{-1}

= 544.44 disintegrations s^{-1}

$\therefore$ Activity $= 544.44$ Bq

Now 1 Ci $= 3.70 \times 10^{10}$ disintegrations s^{-1}

$\therefore$ 544.44 Bq $= 544.44/3.70 \times 10^{10}$ Ci

$= 14.71 \times 10^{-9}$ Ci

$= 14.71$ nCi

Hence (*ii*) is the correct answer.

If your response was (*i*) you will find that to correct counts minute^{-1} to counts second^{-1} you multiplied by 60 rather than dividing by 60.

If your response was (*iii*) you simply forgot to divide by 60 to convert from counts minute^{-1} to counts second^{-1}.

If your response was (*iv*) you forgot to divide by 60 to convert from counts minute^{-1} to counts second^{-1}, *and* you forgot to allow for the 30% efficiency of the counter.

But in every case did you get the conversion Bq $\rightleftharpoons$ Ci correct?

****

SAQ 1.4b

The radioisotope ^{125}I for which $t_{0.5} = 60$ days is widely used in analytical procedures.

Calculate:

(*i*) the activity (in Bq) of 1 g of pure ^{125}I

(*ii*) the mass of a sample of ^{125}I with an activity of 10^3 Bq. $\longrightarrow$

Response

As in the learning text:

$dN/dt = 0.693 \times 6.022 \times 10^{23}/125 \times 60 \times 24 \times 60 \times 60$

$\qquad = 6.44 \times 10^{14}$ Bq (644 TBq)

$\therefore \quad$ Mass of ^{125}I with activity of 10^3 Bq

$\qquad = 10^3/6.44 \times 10^{14}$ g $= 1.553 \times 10^{-12}$ g (1.553 pg)

****

SAQ 1.4c   A sample of the amino acid glycine

$$H_2NCH_2COOH,$$

labelled with ^{14}C, is available at a specific activity of 20 mCi mmol^{-1}. Which of the options (i)–(iii) is the correct value for the specific activity:

(i)   9.86 MBq g^{-1};

(ii)   9.86 TBq g^{-1};

(iii) 9.86 GBq g^{-1}?

Response

20 mCi $= 20 \times 3.7 \times 10^{10} \times 10^{-3}$ disintegration s^{-1} $= 7.4 \times 10^8$ Bq (0.74 GBq)

For glycine, the relative molecular mass $= 75.04$

$\therefore$ 1 mmol $= 0.07504$ g

Hence specific activity $= 7.4 \times 10^8 / 7.504 \times 10^{-2}$

$\qquad = 9.86 \times 10^9$ Bq g^{-1} (9.86 GBq g^{-1})

Thus (*iii*) is the correct answer.

If your response was (*i*) you used moles rather than millimoles in the denominator.

If your response was (*ii*) you used 20 Ci instead of 20 mCi for the activity.

****

SAQ 2.1a   Consideration is being given to the irradiation in a nuclear reactor of each of the following elements. For each element, place a tick in the appropriate column relating to the feasibility of irradiation.

	Definitely Yes	Definitely No	Possibly
Fluorine			
Sodium			
Aluminium			
Phosphorus			
Iodine			
Gold			

Response:

	Definitely Yes	Definitely No	Possibly
Fluorine		√	
Sodium		√	
Aluminium	√		
Phosphorus			√
Iodine		√	
Gold	√		

Fluorine   — definitely no: a highly reactive gas

Sodium   — definitely no: a highly reactive solid

Aluminium — definitely yes: stable, solid metal

Phosphorus — possibly: yellow phosphorus is highly reactive and cannot be used but red phosphorus is stable and may be used

Iodine   — definitely no: likely to be too volatile at reactor temperatures

Gold   — definitely yes: stable, solid metal

****

SAQ 2.1b From SAQ 2.1a you should have appreciated that direct reactor irradiation of sodium metal is unfavoured. From the list of compounds which follow select the most appropriate target materials for preparing radioactive sodium and give your reasons:

$NaCl$; $NaOH$; Na_2O_2; Na_2CO_3; Na_3PO_4.

Response

$NaCl$:   unsuitable—too many other radioactive products.

$NaOH$:   is possible, but has some undesirable properties such as being moisture sensitive

Na_2O_2:   unsuitable—peroxides are too reactive

Na$_2$CO$_3$: the preferred choice—stable, no other products, easily processed

Na$_3$PO$_4$: unsuitable—too many other radioactive products

****

SAQ 2.1c

^{133}BaCl$_2$ solution is available at a specific activity of 370 MBq mg^{-1} Ba. Calculate the disintegration rate of a sample of this solution containing 1 mg of BaCl$_2$, expressing your answer as disintegrations per minute. (A_r(Cl) = 35.45; A_r(Ba) = 137.34).

Response

1 mg Ba $= (137.34 + 2 \times 35.45)/137.34$ mg BaCl$_2$

$= 1.516$ mg BaCl$_2$

370 MBq $= 3.7 \times 10^8$ disintegrations s^{-1}

$= 60 \times 3.7 \times 10^8$ disintegrations min^{-1}

$= 2.22 \times 10^{10}$ disintegrations min^{-1}

$\therefore$ Specific activity of 1 mg of BaCl$_2$

$= 2.22 \times 10^{10}/1.516$

$= 1.464 \times 10^{10}$ disintegrations min^{-1} mg^{-1}

****

SAQ 3.1a   The following isotopes emit the radiation(s) specified:

(*i*) ^{14}C, low energy β^-

(*ii*) ^{60}Co, high energy γ and medium energy β^-

(*iii*) ^{239}Pu, medium energy α

(*iv*) ^{32}P, high energy β^-

Which two of these isotopes can most usefully be measured by proportional counters?

(*i*)   and (*ii*)

(*ii*) and (*iii*)

(*i*)   and (*iii*)

(*iii*) and (*iv*)

(*ii*) and (*iv*)

Which of the isotopes is best counted by end-window Geiger Counting?

Response

(*i*) and (*ii*); ^{60}Co emits high energy, penetrating radiation, so this combination is incorrect.

(*ii*) and (*iii*); incorrect for the same reason.

(*i*) and (*iii*); both isotopes emit radiation of low penetrating power, so this is the correct answer.

(*iii*) and (*iv*); ^{32}P emits high energy, quite penetrating radiation, so this combination is incorrect.

(*ii*) and (*iv*); incorrect for the same reason.

The isotope *best* suited to end-window Geiger counting is ^{32}P.

The penetration of ^{14}C and ^{239}Pu radiation is too low. Note the emphasis on the word 'best'. It would be possible to count ^{60}Co in a Geiger counter, because of the medium energy β^- radiation, but the high energy radiation is more efficiently counted by scintillation methods (see 3.2).

****

SAQ 3.2a   The following is a list of key words and compounds from this section. Place them in the correct pairings.

thallium(I) iodide; secondary solute; dynode; magnesium oxide; anthracene; caesium film; lattice dislocation; beryllium alloy; primary solute; photocathode; 1,4-di-2-(5-phenyloxazole); p-terphenyl; reflective layer; organic crystal.

Response

thallium(I) iodide   and   lattice dislocation

magnesium oxide   and   reflective layer

anthracene   and   organic crystal

caesium film   and   photocathode

p-terphenyl and primary solute

beryllium alloy and dynode

1,4-di-2-(5-phenyloxazole) and secondary solute

If you are unsure of any of the pairings re-read Section 3.2.

****

SAQ 3.3a The table below gives E_γ values (keV) for four isotopes. Calculate ΔE values, and complete the table. You should use resolutions of 8% and 0.3% for NaI(Tl) and Ge(Li) respectively. From your calculated ΔE values, decide which *two* pairs of gamma rays could *not* be resolved satisfactorily by a NaI(Tl) detector. Can they be resolved satisfactorily by a Ge(Li) detector?

		ΔE/keV	
Isotope	E_γ/keV	Ge(Li)	NaI(Tl)
^{198}Au	412		
^{69}Zn	440		
^{76}As	559		
^{122}Sb	564		$\longrightarrow$

Response

$$\Delta E/\text{keV}$$

Isotope	E_γ/keV	Ge(Li)	NaI(Tl)
^{198}Au	412	1.24	32.96
^{69}Zn	440	1.32	35.20
^{76}As	559	1.68	44.72
^{122}Sb	564	1.69	45.12

It should be obvious to you that the γ ray energies for ^{76}As and ^{122}Sb are so close that the photopeaks from a NaI(Tl) detector are bound to overlap: the E_γ values are only 5 keV apart and the ΔE values are $\simeq$ 45 keV.

If we then consider the photopeaks for $E_\gamma = 412$ and 440 keV respectively, and assume they are symmetrical about the mid-point, then $E + \Delta E/2$ for $E = 412$ is $412 + 16.48$ ie 428.48, and $E - \Delta E/2$ for $E = 440$ is $440 - 17.60$ ie 422.40. Hence these two peaks will overlap also.

By a similar argument you can show that a Ge(Li) detector will not only resolve peaks at 412 and 440 keV, but will also resolve peaks at 559 and 564 keV.

$$559 + 1.68/2 = 559.84$$

hence no overlap

$$564 - 1.69/2 = 563.16$$

****

SAQ 3.5a	In an experiment to check that a Geiger counter was working the following 20 sets of counts were obtained, in each case over a 30 second period.

9843, 9774, 9858, 9828, 9831, 9768, 9927, 9834, 9792, 9804, 9819, 9882, 9846, 9816, 9696, 9897, 9687, 9876, 9789, 9858.

(*i*) Calculate the mean value of these results, $\bar{x}$.

(*ii*) Obtain an estimate of the true standard deviation, s.

(*iii*) Calculate the relative standard deviation (coefficient of variation).

(*iv*) Check how many individual results lie in the range $(\bar{x} - s)$ to $(\bar{x} + s)$.

Response

(*i*) The mean value is the sum of all the results, divided by the number of results (20).

Thus $\bar{x} = 9821$

(*ii*) The most widely used formula for estimating the standard deviation is:

$$s \quad \text{(sometimes written as } \sigma) = \left[\frac{\sum\limits_{i=1}^{i=N} (x_i - \bar{x})^2}{N - 1} \right]^{\frac{1}{2}}$$

where N is the number of results, and $(x_i - \bar{x})$ is the difference of each individual result from the mean.

Thus $s = (68705/19)^{1/2}$

$= 60.13$

(*iii*) The relative standard deviation (RSD)

$= s \times 100/\bar{x}$

$= 60.13 \times 100/9821$

$= 0.612$

(*iv*) The range $(\bar{x} - s)$ to $(\bar{x} + s)$ is 9761–9881. Thus 15 of the 20 results (75%) lie in this range.

****

SAQ 3.5b	Preliminary experiments show that the count rate of a sample is 4860 counts min^{-1} and that of the background is 60 counts min^{-1}. You have 20 minutes available for counting both sample and background for periods which optimise the statistics for each. What are the best values of t_s and t_b?

Response

$t_s/t_b = \sqrt{4860/60}$

$= \sqrt{81}$

$= 9$

$\therefore t_s = 9t_b$

Since $t_s + t_b = 20$

$9t_b + t_b \qquad = 20$

$\therefore \quad t_b \qquad = 2$

Hence $t_s \qquad = 18$ minutes

****

SAQ 3.6a The dose rate at the surface of a ^{60}Co γ ray source is 160 mrem h^{-1}. Given that the half-thickness value for lead is 1.25 cm, what thickness of lead shielding is necesssary to reduce the dose rate to 10 mrem h^{-1}?

Response

To reduce the dose rate from 160 to 10 mrem h^{-1} means a reduction by a factor of 16, ie it has to be halved four times:

$$160 \rightarrow 80 \rightarrow 40 \rightarrow 20 \rightarrow 10$$

We need four 'half-thicknesses' of lead, ie 4×1.25 cm

$$= 5 \text{ cm of lead.}$$

****

SAQ 3.6b   Given that for ^{60}Co the absorption coefficient of
                concrete is $\mu = 0.075$ cm^{-1}, calculate the thick-
                ness of concrete necessary to achieve the reduc-
                tion of dose rate from 160 to 10 mrem h^{-1}.

Response

The equation is $N = N_0 \, e^{-\mu x}$

$N = 10, \, N_0 = 160, \qquad \mu = 0.075$ cm^{-1}

$\therefore \quad 10 = 160 \, e^{-0.075x}$

$\therefore \quad 1/16 = e^{-0.075x}$

$\therefore \quad 16 = e^{0.075x}$

$\therefore \quad x = 37$ cm

****

SAQ 3.6c   Calculate the dose rate in SI units for the fol-
                lowing 0.37 GBq sources and distances:

                (i) a β^- source at 10 cm;

                (ii) a γ source of total energy 2.5 MeV at 1 m.

Response

(i) For the β^- source the correct equation is 3.6d:

Dose rate $= 0.81 \times 0.37$

$= 0.30$ Gy h^{-1} at 10 cm

(*ii*) For the γ source the correct equation is 3.6b:

Dose rate $= 143 \times 0.37 \times 2.5$

$= 132.3$ μGy h^{-1} at 1 metre

****

SAQ 3.6d In SAQ 3.6c part (*ii*), you calculated the γ dose rate at 1 metre. Will the γ dose rate at 10 cm be greater or lesser than that for the β emitter you calculated in part (*i*)?

Response

If the dose at 100 cm is 132.3 μGy h^{-1} the dose at 10 cm will be 10^2 greater according to the inverse square law.

$\therefore$ It will be 1.323×10^4 μGy h^{-1} (0.01323 Gy h^{-1}) which is less than that for the β emitter.

****

SAQ 4.1a	Assuming that a liquid Geiger counter has an efficiency of 30% for ^{40}K radiation, calculate the expected count rate for a 10 cm^3 sample of KCl solution containing 50 g dm^{-3} KCl. [A_r(K) = 39.1, A_r(Cl) = 35.45, specific activity = 1.85 × 10^3 disintegrations min^{-1} g^{-1} potassium].

Response

It is necessary initially to calculate the potassium content of 10 cm^3 of a solution of the given concentration.

This is 10 × 50 × 39.1/1000(39.1 + 35.45) g

= 0.262 g

The specific activity = 1.85 × 10^3 disintegrations min^{-1} g^{-1} of potassium.

∴   For this sample the theoretical activity

= 1.85 × 10^3 × 0.262

= 485 disintegrations min^{-1}

Since the counter is 30% efficient this is 0.3 × 485 counts min^{-1}

= 145.5 counts min^{-1}

****

SAQ 4.1b	All naturally occurring rubidium ores contain ^{87}Sr resulting from β^- decay of ^{87}Rb. In naturally occurring rubidium, 278 of every 1000 rubidium atoms are ^{87}Rb. A mineral containing 0.85% rubidium was analysed and found to contain 0.0098% strontium. Assuming that all this strontium originated from decay of ^{87}Rb, estimate the age of the mineral.

$(t_{0.5}$ of $^{87}Rb = 6.2 \times 10^{10}$ years$)$.

Response

Basically you need to calculate the amount of ^{87}Rb now, and the amount originally, and then use the simple decay law

Amount now $(N) = 0.278 \times 0.85 = 0.2362$

Amount originally $(N_0) = 0.2362 + 0.0098 = 0.246$

$\therefore \quad N/N_0 = e^{-\lambda t}$

$\therefore \quad 0.236/0.246 = e^{-\lambda t}$

$\therefore \quad \ln 0.236 - \ln 0.246 = -\lambda t$

and $\lambda = 0.693/t_{0.5} = 0.693/6.2 \times 10^{10}$ years^{-1}

$\therefore \quad \ln 0.236 - \ln 0.246 = -0.693t/6.2 \times 10^{10}$

$\therefore \quad -1.444 + 1.402 = -1.118 \times 10^{-11} \times t$

$\therefore \quad t = 0.042/1.118 \times 10^{-11}$

$\therefore \quad t = 3.67 \times 10^9$ years

****

SAQ 4.2a The filtrate from a gravimetric procedure for phosphate, labelled with ^{32}P was made up to a volume of 100 cm^3. An aliquot of 10 cm^3 gave 1750 counts when counted for 5 minutes in a liquid Geiger counter. The precipitate from the procedure was dissolved in exactly 10 cm^3 of solution, and gave 14 500 counts min^{-1} when counted under identical conditions. The efficiency of precipitation is:

A 97.64%

B 80.55%

C 45.31%

D 89.23%

Response

The count rates must be in the same units—counts min^{-1}

∴   for filtrate, 10 cm^3 gives 350 counts min^{-1}

The total volume of filtrate was 100 cm^3

∴   for total filtrate, count rate = 3500 counts min^{-1}

∴   efficiency of precipitation

= [14 500 × 100/(14 500 + 3500)]

= 80.55%

Answer B is correct

If your answer was A, you forgot that the filtrate was diluted to 100 cm^3, and have calculated efficiency as: $14\,500 \times 100/(14\,500 + 350)$.

If your answer was C, you forgot that the filtrate was counted for five minutes, and have calculated efficiency as $14\,500 \times 100/(14\,500 + 17\,500)$.

If your answer was D, you forgot that the filtrate was counted for five minutes and also that it was diluted to 100 cm^3, so you calculated efficiency as $14\,500 \times 100/(14\,500 + 1750)$.

****

SAQ 4.2b

In the nuclear industry, uranium in the form of the UO_2^{2+} ion is extracted from an aqueous to an organic phase using the complexing agent tri-n-butylphosphate (TBP) dissolved in kerosene. In order to carry out a laboratory experiment to investigate the efficiency of this separation an aqueous solution containing $UO_2(NO_3)_2$ was shaken with an equal volume of TBP/kerosene. After the phases had been separated 10 cm^3 of each phase was counted in a liquid Geiger counter. The organic phase gave 11 300 counts min^{-1} and the aqueous phase gave 4394 counts min^{-1}.

Calculate:

(*i*) the % efficiency of the separation;

(*ii*) the number of extractions necessary to extract >99% of the uranium into the organic phase. $\longrightarrow$

Response

(*i*) Since equal volumes are involved the % efficiency is simply the ratio of the count in the organic phase to the *total* count:

Efficiency = 11 300 × 100/(11 300 + 4394)

= 72.00%

(*ii*) This would leave 28% uranium in the aqueous phase. A second extraction would remove 0.72 × 28 = 20.16%

Thus the total so far extracted = (72.00 + 20.16) = 92.16%

This would leave 7.84% uranium in the aqueous phase.

A third extraction would remove 0.72 × 7.84 = 5.64%.

Thus the total so far extracted = (92.16 + 5.64) = 97.80%

This would leave 2.20% uranium in the aqueous phase.

A fourth extraction would remove 0.72 × 2.20 = 1.58%

Thus the total so far extracted = (97.80 + 1.58) = 99.38%

∴   Four extractions are needed to remove >99%.

****

SAQ 4.3a | It is required to measure the orthophosphate content of a solution X. A 1 cm³ sample of X, of density 1 g cm⁻³, was taken, and 3.0 mg of ^{32}P-labelled PO_4^{3-} of specific activity 0.09 μCi mg⁻¹ was added. From this mixture a pure sample of orthophosphate was isolated; the sample weighed 30 mg, and was found to give 1.8×10^4 counts min⁻¹ in a counter of 30% detection efficiency. Calculate the concentration of orthophosphate in X, expressing your answer as % w/w.

Response

It is important to be consistent in the use of units.

The count rate of the sample $= 1.8 \times 10^4$ counts min⁻¹

$\therefore$ Disintegration rate (counter 30% efficient)

$= 1.8 \times 10^4 \times 100/30$

$= 6 \times 10^4$ disintegrations min⁻¹

Since 1 μCi $= 2.22 \times 10^6$ disintegrations min⁻¹ the activity isolated

$= 6 \times 10^4/2.22 \times 10^6 \ \mu$Ci

$= 2.703 \times 10^{-2} \ \mu$Ci

And the new specific activity $= 2.703 \times 10^{-2}/30 \ \mu$Ci mg⁻¹

$= 9 \times 10^{-4} \ \mu$Ci mg⁻¹

Now $w_x = w_0[(S_0/S_1) - 1]$ (4.3a)

$$= 3.0[(0.09/9 \times 10^{-4}) - 1]$$

$$= 3.0(10^2 - 1)$$

$$= 297 \text{ mg}$$

The original solution was a 1 cm^3 aliquot which had a density of 1 g cm^{-3} ie 1000 mg cm^{-3}

$\therefore$ concentration of orthophosphate $= 297 \times 100/1000$

$$= 29.7\% \text{ w/w}$$

****

SAQ 4.3b	In order to set up an RIA procedure for measuring insulin a calibration curve was first prepared. Standard volumes of insulin solution of known concentration were mixed with a fixed volume of labelled insulin, such that the final concentration of insulin in each sample was 3, 5, 7 and 9 ng cm^{-3} respectively. The total activity of each solution was 2×10^4 counts min^{-1}. The same quantity of antibody was added to each solution, the mixture equilibrated, and in each case the insulin–antibody complex was isolated and the activity measured. The same procedure was then followed using a known volume of a solution containing an unknown amount of insulin.

The results are tabulated.

Concentration of insulin/ng cm^{-3}	3.0	5.0	7.0	9.0	Unknown
Activity of bound complex/counts min^{-1}	13 245	11 111	9 852	9 091	10 100

$\longrightarrow$

SAQ 4.3b (cont.)	Draw an appropriate calibration curve, and hence calculate the concentration of insulin /ng cm^{-3} in the 'unknown' solution for which the count was taken.

Response

Since you know that the total count for each solution was 20 000 counts min^{-1} and you are told the count rate for the bound insulin you can by difference calculate the count rate for the unbound insulin, and hence calculate $R^*_{f/b}$:

Concn. /ng cm^{-3}	3.0	5.0	7.0	9.0
Total count	20 000	20 000	20 000	20 000
Bound	13 245	11 111	9852	9091
Unbound (Free)	6755	8889	10 148	10 909
$R^*_{f/b}$	0.57	0.80	1.03	1.20

Hence you can plot the calibration curve of $R^*_{f/b}$ against concentration (Fig. 4.3b). Note that the amount of unbound labelled insulin necessarily increases as the amount of inactive insulin increases.

For the unknown, the unbound insulin $= (20\,000 - 10\,100) = 9900$ counts min^{-1} so that $R^*_{f/b} = 9900/10\,100 = 0.98$. By reference to the calibration curve you will see that this corresponds to an insulin concentration of 6.5 ng cm^{-3} in the solution as measured.

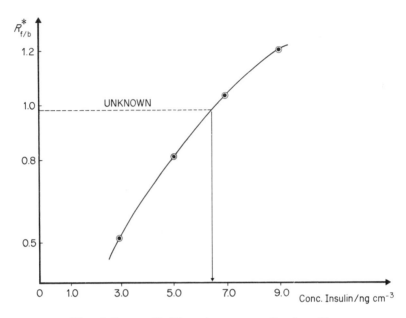

Fig. 4.3c.   *Calibration curve for insulin*

Note: An alternative method is to calculate the % bound activity and plot this against concentration:

Concn./ng cm^{-3}	3.0	5.0	7.0	9.0
Total count	20 000	20 000	20 000	20 000
Bound	13 245	11 111	9852	9091
% bound	66.2	55.6	49.3	45.5

For the unknown, % bound = 10 100 × 10^2/20 000 = 50.5%, and reading this off the appropriate curve again gives 6.5 ng cm^{-3} in the final solution of unknown.

****

SAQ 4.4a	A 1 g sample of an alloy containing a small quantity of gold was irradiated and counted under identical conditions to a standard containing 10 mg of gold. The unknown gave an activity of 500 disintegrations min^{-1}, and the standard gave 2000 disintegrations min^{-1}. Calculate the % (w/w) of gold in the alloy.

Response

$$A_{standard} = 2000 \text{ disintegrations } min^{-1}$$

$$A_{unknown} = 500 \text{ disintegrations } min^{-1}$$

$$w_{standard} = 10 \text{ mg}$$

$$\therefore \quad 2000/500 = 10/w_{unknown}$$

$$\therefore \quad w_{unknown} = 5000/2000 = 2.5 \text{ mg}$$

The alloy weighed 1 g (1000 mg) thus

$$\% \text{ gold (w/w)} = 2.5 \times 100/1000$$

$$= 0.25\%$$

****

SAQ 4.4b

Consider the determination of manganese by neutron activation analysis. The only naturally occurring isotope of manganese, $^{55}_{25}Mn$, has a half-life of 2.58 h, and a capture cross-section of 13.3 barn. Calculate the theoretical count rate (counts $min^{-1} kg^{-1}$) in a detector which has an efficiency of 10% under the following experimental conditions: neutron flux = 5×10^{17} neutrons $s^{-1} m^{-2}$; time of irradiation = $6 \times t_{0.5}$ and the precounting delay time = $0.5 t_{0.5}$. Which of the following answers is correct?

(*i*) 1.076×10^{16};

(*ii*) 3.044×10^{16};

(*iii*) 2.899×10^{16};

(*iv*) 3.044×10^{17};

(*v*) 8.457×10^{12};

Response

Using Eq. 4.4c:

$$S = \frac{6.022 \times 10^{26} \times 5 \times 10^{17} \times 13.3 \times 10^{-28} \times 0.5^{0.5}(1 - 0.5^6)}{54.94}$$

$= 5.074 \times 10^{15}$ disintegrations $s^{-1} kg^{-1}$

$= 5.074 \times 10^{14}$ counts $s^{-1} kg^{-1}$ (10% efficiency)

$= 3.044 \times 10^{16}$ counts $min^{-1} kg^{-1}$

Hence (*ii*) is the correct answer.

If your answer was (*i*) you put $(0.5)^{T/t_{0.5}} = 0.5^2$ instead of $0.5^{0.5}$ which is easy to do!

If you answered (*iii*) you used $t = 4t_{0.5}$ (as in the original example) rather than $t = 6t_{0.5}$ as specified.

If your answer was (*iv*) you forgot to allow for the counter being 10% efficient.

Finally if your answer was (*v*) you made the surprisingly common error of dividing by 60 instead of multiplying by 60 when converting seconds to minutes!

****

SAQ 4.4c Assuming that a neutron source is sufficiently small to be considered as a 'point source', and by calculating the surface area of a sphere of 1 cm radius, calculate the neutron flux at 1 cm for a 3 Ci source of ^{241}Am/Be.

Response

For a 3 Ci source the output is

$$3 \times 2.2 \times 10^6 \text{ neutrons s}^{-1}$$

The surface area of a sphere is $4\pi r^2$

Hence when $r = 1$ cm, the surface area is 4π cm^2

$\therefore$ Flux at 1 cm $= 6.6 \times 10^6/4\pi$

$$= 6.6 \times 10^6/4 \times 3.142$$

$$= 5.25 \times 10^5 \text{ neutrons s}^{-1} \text{ cm}^{-2}$$

$$= 5.25 \times 10^9 \text{ neutrons s}^{-1} \text{ m}^{-2}.$$

****

SAQ 5.1a

The relevant data for the determination given in the paper are:

σ/m^2 for $^{23}\text{Na} \rightarrow {}^{24}\text{Na}$, $^{75}\text{As} \rightarrow {}^{76}\text{As}$, and $^{121}\text{Sb} \rightarrow {}^{122}\text{Sb}$

= 0.53 $\times$ 10^{-28}, 4.30 $\times$ 10^{-28}, and 3.90 $\times$ 10^{-28} respectively.

A for Na, As, and Sb are 22.99, 74.92, and 121.75 g respectively.

f for ^{23}Na, ^{75}As and ^{121}Sb are 1.0, 1.0, and 0.573 respectively.

Using the values of ϕ, $t_{0.5}$, and irradiation time given in the paper and taking $T = 4$ days, calculate the relative activities for a sample of detergent containing 10^4 ppm Na, 25 ppm As, and 5 ppm Sb.

Response

This is a demanding SAQ—certainly the hardest you have tackled so far. Let us calculate the induced specific activity for each element in turn.

Sodium

Eq. 4.4c is used here.

$$(0.5)^{T/t_{0.5}} = (0.5)^{4 \times 24/15.4} = 0.013$$

$$(0.5)^{t/t_{0.5}} = (0.5)^{20/15.4 \times 60} = 0.985$$

Thus $[1 - (0.5)^{t/t_{0.5}}] = 1 - 0.985 = 0.015$

Note that the units of T and $t_{0.5}$ are the same, and the units of t and $t_{0.5}$ are the same, so that in each case they are consistent.

Substituting these values and other information in Eq. 4.4c gives:

$$S = \frac{6.022 \times 10^{26} \times 0.53 \times 10^{-28} \times 10^{17} \times 1 \times 0.013 \times 0.015}{22.99}$$

$$= 2.707 \times 10^{10} \text{ disintegrations s}^{-1} \text{ kg}^{-1}$$

Arsenic

$$0.5^{96/26.4} = 0.0804$$

$$0.5^{20/1584} = 0.9913$$

$$S = \frac{6.022 \times 10^{26} \times 4.30 \times 10^{-28} \times 10^{17} \times 1 \times 0.0804 \times 0.0087}{74.92}$$

$$= 2.418 \times 10^{11} \text{ disintegrations s}^{-1} \text{ kg}^{-1}.$$

Antimony

$$0.5^{4/2.7} = 0.358$$

$$0.5^{20/3888} = 0.996$$

$$S = \frac{6.022 \times 10^{26} \times 3.90 \times 10^{-28} \times 10^{17} \times 0.573 \times 0.358 \times 0.004}{121.75}$$

$$= 1.583 \times 10^{11} \text{ disintegrations s}^{-1} \text{ kg}^{-1}$$

Now the relative concentrations of Na : As : Sb are $10^4 : 25 : 5$ so that the relative activities are:

$2.707 \times 10^{10} \times 10^4 : 2.418 \times 10^{11} \times 25 : 1.583 \times 10^{11} \times 5$

ie $2.707 \times 10^{14} = 6.045 \times 10^{12} : 7.915 \times 10^{11}$

Hence the ratio is roughly $342 : 7.6 : 1$

In a sense the exact value of the ratio doesn't matter; the point is that the activity from Na is far higher than the other two. This is particularly true when you realise that the concentration of sodium quoted (10^4 ppm; ie 1% w/w) is probably low for a material of this kind.

****

SAQ 5.1b   A standard containing 3 μg of arsenic gave 1200 counts min^{-1}, and a sample of the unknown weighing 0.4955 g gave 2081 counts min^{-1}. Calculate the arsenic content in the unknown.

Response

Using the equation given in the paper (p. 457):

Concentration of arsenic $= \dfrac{3 \times 2081}{0.4955 \times 1200}$

$\therefore$ Arsenic $= 10.5$ ppm

****

SAQ 5.1c	The paper states that the specific activity of the iodinated barbituric acid is 1.56 TBq mmol^{-1} or 4.18 MBq μg^{-1}. What is the molar mass of the compound expressed in grams?

Response

In the first set of units

Specific Activity

$= 1.56 \times 10^{12}/10^{-3}$ M disintegrations s^{-1} g^{-1}

In the second set of units

Specific Activity

$= 4.18 \times 10^{6}/10^{-6}$ disintegrations s^{-1} g^{-1}

Thus $1.56 \times 10^{12}/10^{-3}$ M $= 4.18 \times 10^{6}/10^{-6}$

ie   M $= 1.56 \times 10^{12} \times 10^{-6}/10^{-3} \times 4.18 \times 10^{6}$ g

$\therefore$ M $= 373.2$ g

****

SAQ 5.1d	Given that the relative atomic mass of rhenium is 186.21, calculate the concentration of Re in ppm of a solution of ReO$_4^-$ of concentration 5 $\times$ 10^{-5} mol dm^{-3}. $\longrightarrow$

Response

The solution is 5×10^{-5} mol dm^{-3} in ReO_4^-, and since there is one Re atom per ReO_4^- it is also 5×10^{-5} mol dm^{-3} in Re.

Weight concentration is molarity $\times$ relative atomic mass

$\therefore$ Concentration of Re $= 5 \times 10^{-5} \times 186.21$ g dm^{-3}

$\qquad = 931 \times 10^{-5}$ g dm^{-3}

$\qquad = 9.31 \times 10^{-3}$ g dm^{-3}

$\therefore$ In 10^6 cm^3 (ie 10^6 g sample) there would be 9.31 g Re

ie 9.31 parts per million

Units of Measurement

For historic reasons a number of different units of measurement have evolved to express quantity of the same thing. In the 1960s, many international scientific bodies recommended the standardisation of names and symbols and the adoption universally of a coherent set of units—the SI units (Système Internationale d'Unités)—based on the definition of five basic units: metre (m); kilogram (kg); second (s); ampere (A); mole (mol); and candela (cd).

The earlier literature references and some of the older text books, naturally use the older units. Even now many practicing scientists have not adopted the SI unit as their working unit. It is therefore necessary to know of the older units and be able to interconvert with SI units.

In this series of texts SI units are used as standard practice. However in areas of activity where their use has not become general practice, eg biologically based laboratories, the earlier defined units are used. This is explained in the study guide to each unit.

Table 1 shows some symbols and abbreviations commonly used in analytical chemistry; Table 5 is a glossary of abbreviations used in this particular text. Table 2 shows some of the alternative methods for expressing the values of physical quantities and the relationship to the value in SI units.

More details and definition of other units may be found in the *Manual of Symbols and Terminology for Physicochemical Quantities and Units*, Whiffen, 1979, Pergamon Press.

Table 1 *Symbols and Abbreviations Commonly used in Analytical Chemistry*

Å	Angstrom
$A_r(X)$	relative atomic mass of X
A	ampere
E or U	energy
G	Gibbs free energy (function)
H	enthalpy
J	joule
K	kelvin ($273.15 + t\,°C$)
K	equilibrium constant (with subscripts p, c, therm etc.)
K_a, K_b	acid and base ionisation constants
$M_r(X)$	relative molecular mass of X
N	newton (SI unit of force)
P	total pressure
s	standard deviation
T	temperature/K
V	volume
V	volt ($J\ A^{-1}\ s^{-1}$)
$a, a(A)$	activity, activity of A
c	concentration/ mol dm^{-3}
e	electron
g	gramme
i	current
s	second
t	temperature / °C
bp	boiling point
fp	freezing point
mp	melting point
$\approx$	approximately equal to
$<$	less than
$>$	greater than
e, $\exp(x)$	exponential of x
ln x	natural logarithm of x; ln $x = 2.303 \log x$
log x	common logarithm of x to base 10

Table 2 *Alternative Methods of Expressing Various Physical Quantities*

1. **Mass (SI unit : kg)**

    $$g = 10^{-3} \text{ kg}$$
    $$mg = 10^{-3} \text{ g} = 10^{-6} \text{ kg}$$
    $$\mu g = 10^{-6} \text{ g} = 10^{-9} \text{ kg}$$

2. **Length (SI unit : m)**

    $$cm = 10^{-2} \text{ m}$$
    $$\text{Å} = 10^{-10} \text{ m}$$
    $$nm = 10^{-9} \text{ m} = 10\text{Å}$$
    $$pm = 10^{-12} \text{ m} = 10^{-2} \text{ Å}$$

3. **Volume (SI unit : m³)**

    $$l = dm^3 = 10^{-3} \text{ m}^3$$
    $$ml = cm^3 = 10^{-6} \text{ m}^3$$
    $$\mu l = 10^{-3} \text{ cm}^3$$

4. **Concentration (SI units : mol m^{-3})**

    $$M = \text{mol l}^{-1} = \text{mol dm}^{-3} = 10^3 \text{ mol m}^{-3}$$
    $$\text{mg l}^{-1} = \mu g \text{ cm}^{-3} = ppm = 10^{-3} \text{ g dm}^{-3}$$
    $$\mu g \text{ g}^{-1} = ppm = 10^{-6} \text{ g g}^{-1}$$
    $$\text{ng cm}^{-3} = 10^{-6} \text{ g dm}^{-3}$$
    $$\text{ng dm}^{-3} = \text{pg cm}^{-3}$$
    $$\text{pg g}^{-1} = ppb = 10^{-12} \text{ g g}^{-1}$$
    $$mg\% = 10^{-2} \text{ g dm}^{-3}$$
    $$\mu g\% = 10^{-5} \text{ g dm}^{-3}$$

5. **Pressure (SI unit : N m^{-2} = kg m^{-1} s^{-2})**

    $$Pa = Nm^{-2}$$
    $$atmos = 101\ 325 \text{ N m}^{-2}$$
    $$bar = 10^5 \text{ N m}^{-2}$$
    $$torr = mmHg = 133.322 \text{ N m}^{-2}$$

6. **Energy (SI unit : J = kg m² s^{-2})**

    $$cal = 4.184 \text{ J}$$
    $$erg = 10^{-7} \text{ J}$$
    $$eV = 1.602 \times 10^{-19} \text{ J}$$

Table 3   *Prefixes for SI Units*

Fraction	Prefix	Symbol
10^{-1}	deci	d
10^{-2}	centi	c
10^{-3}	milli	m
10^{-6}	micro	μ
10^{-9}	nano	n
10^{-12}	pico	p
10^{-15}	femto	f
10^{-18}	atto	a

Multiple	Prefix	Symbol
10	deka	da
10^2	hecto	h
10^3	kilo	k
10^6	mega	M
10^9	giga	G
10^{12}	tera	T
10^{15}	peta	P
10^{18}	exa	E

Table 4   *Recommended Values of Physical Constants*

Physical constant	Symbol	Value
acceleration due to gravity	g	9.81 m s^{-2}
Avogadro constant	N_A	$6.022\ 05 \times 10^{23}$ mol^{-1}
Boltzmann constant	k	$1.380\ 66 \times 10^{-23}$ J K^{-1}
charge to mass ratio	e/m	$1.758\ 796 \times 10^{11}$ C kg^{-1}
electronic charge	e	$1.602\ 19 \times 10^{-19}$ C
Faraday constant	F	$9.648\ 46 \times 10^{4}$ C mol^{-1}
gas constant	R	8.314 J K^{-1} mol^{-1}
'ice-point' temperature	T_{ice}	273.150 K exactly
molar volume of ideal gas (stp)	V_m	$2.241\ 38 \times 10^{-2}$ m^3 mol^{-1}
permittivity of a vacuum	ϵ_o	$8.854\ 188 \times 10^{-12}$ kg^{-1} m^{-3} s^4 A^2 (F m^{-1})
Planck constant	h	$6.626\ 2 \times 10^{-34}$ J s
standard atmosphere pressure	p	$101\ 325$ N m^{-2} exactly
atomic mass unit	m_u	$1.660\ 566 \times 10^{-27}$ kg
speed of light in a vacuum	c	$2.997\ 925 \times 10^{8}$ m s^{-1}

Table 5 *Glossary and Abbreviations used in Radiochemistry*

Bq   becquerel—SI unit of radiation activity
$$(= 27 \text{ pCi})$$

Ci   curie—pre-SI unit of radiation activity
$$(= 3.7 \times 10^{10} \text{ Bq})$$

Gy   gray—derived SI unit of absorbed radiation
$$(1 \text{ J kg}^{-1} = 100 \text{ rad})$$

R   roentgen—pre-SI unit of exposure dose
$$(2.58 \times 10^{-4} \text{ C kg}^{-1})$$

Sv   sievert—derived SI unit of dose equivalent
$$(= 1 \text{ J kg}^{-1})$$

λ decay constant

$t_{0.5}$ half life

rad   pre-SI unit of absorbed dose
$$(1 \text{ rad} = 10^{-2} \text{ J kg}^{-1})$$

A mass number

N number of neutrons

Z atomic number

$_{-1}^{0}e$, e   electron

$_{0}^{1}n$, n   neutron

$_{1}^{1}p$, p   proton

f fractional abundance of a particular isotope

ϕ particle flux

σ capture cross-section

$_{+1}^{0}\beta$ positron